D0147059

The Insecure Alliance

THE INSECURE ALLIANCE

Energy Crises and Western Politics Since 1944

ETHAN B. KAPSTEIN

Written under the auspices
of the Center for International Affairs
Harvard University

New York Oxford
OXFORD UNIVERSITY PRESS
1990

Oxford University Press

Oxford New York Toronto
Delhi Bombay Calcutta Madras Karachi
Petaling Jaya Singapore Hong Kong Tokyo
Nairobi Dar es Salaam Cape Town
Melbourne Auckland

and associated companies in
Berlin Ibadan

Published by Oxford University Press, Inc.,
200 Madison Avenue, New York, New York 10016

Library of Congress Cataloging-in-Publication Data
Kapstein, Ethan B.
The insecure alliance :
energy crises and western politics since 1944 /
Ethan B. Kapstein.
p. cm.
"Written under the auspices
of the Center for International Affairs, Harvard University."
Bibliography: p. Includes index. ISBN 0-19-505851-8
1. Energy industries—Political aspects—History—20th century.
2. World politics—1945- I. Harvard University.
Center for International Affairs. II. Title.
HD9502.A2K375 1990 333.79'09'045—dc20 89-32818 CIP

9 8 7 6 5 4 3 2 1

Printed in the United States of America
on acid-free paper

To My Family

Preface

This book examines Western alliance relations during every major energy crisis since World War II. As I will show, public officials did not view these shortages merely as technical issues of balancing supply and demand but as complex political crises that threatened highly valued national interests. Because energy security was deemed vital to economic well-being and military preparedness, the allies attempted to formulate a collective response to each crisis. Their efforts, however, did not always succeed.

My objectives in the study are threefold. First, I seek to explain why the allies were able to coordinate their policy response to some energy crises but not to others. Second, I attempt to show why the search for alliance energy security has proved so elusive. Third, I argue that economic and security issues have been inextricably linked in the Western alliance.

Not surprisingly, the United States served as the prime mover in promoting alliance cooperation in the energy issue area. But few works in the now vast literature on energy policy highlight this alliance dimension. As I hope the book will make clear, the development of U.S. energy policy did not take place within a political context that defined security and prosperity solely in national terms, but more broadly with regard to the requirements of the anti-Soviet allies. Scholars have already pointed out the difficulty the executive branch faced after World War II in formulating policies that took into account the interests of international oil companies, domestic energy producers, and American consumers. But few have taken into account the additional problems posed by alliance energy needs.

This study of alliance energy relations is located at the interface of scholarship in international security and international political economy. Since the end of World War II, U.S. policymakers have recognized that economically vibrant allies were vital to the West's military preparedness and political stability. Indeed, Article II of the North Atlantic Treaty (signed in April 1949 creating NATO) pledged that "the Parties . . . will seek to eliminate conflict in their international economic policies and will encourage collaboration between any and all of them." To the postwar statesmen who had experienced war and depression, it was evident that economic well-being and national security were interdependent.

Energy, however, has long been perceived as the allies' Achilles' heel. "Perhaps the most vulnerable sector of NATO's western European members," Ronald Ritchie wrote in his 1956 book, *NATO: The Economics of an Alliance,* "lies in the field of fuel and power requirements. . . . Here is one of NATO's exposed flanks, equally vulnerable whether the battle lines are drawn for hot war, cold war, or competitive coexistence." Since the war's end the West has been dependent on one unreliable energy source after another, including European coal, and oil from the Middle East, North Africa, and the Soviet Union. The West has been rocked by seven major fuel crises, beginning in the years 1944, 1951, 1956, 1967, 1973, 1978, and 1980. Energy security has thus been a "high policy" concern of alliance officials throughout the postwar period.

The story of these energy shortages is presented in nine chapters. The first chapter builds a conceptual framework for analyzing alliance relations during energy crises, drawing heavily on hegemonic stability theory and bargaining theory. Chapter 2 and Chapters 4 through 8 provide detailed historical case studies of the crises caused by Europe's postwar coal shortage, the Iranian nationalization of British oil assets in 1951, the nationalization of the Suez Canal Company in 1956, the oil embargoes of 1967 and 1973, the Iranian revolution of 1978–1979, and the outbreak of the Iran–Iraq war in September 1980. Chapter 3 describes western Europe's transition during the Marshall Plan years from an economy based on indigenous coal to one based on imported oil, and U.S. policies that were meant to ensure energy security despite the shift. Chapter 9 provides conclusions.

I began the research for this book over ten years ago, between the Arab oil embargo of 1973–1974 and the Iranian revolution of 1978–1979. At that time, there was widespread concern about the strength and cohesion of the Western alliance. I thought that historical analysis might provide us with some insights regarding the links between economic well-being and Western security, and the "energy crisis" provided an obvious opening into this broader issue. Most of the historical work was conducted in the National Archives in Washington, the Harry S. Truman Library in Independence, Missouri, and the archives of the Organization for Economic Cooperation and Development (OECD) and the International Energy Agency (IEA) in Paris, France. It is unfortunate, but under OECD regulations I have not been granted permission to publish specific citations to the OECD and IEA documents used in the study. Readers who seek detailed references may consult my unpublished doctoral dissertation, "American Strategy and Alliance Energy Security," which is located at the library of the Fletcher School of Law and Diplomacy, Medford, Massachusetts. I cannot, of course, guarantee that other scholars will be able to gain access to the OECD and IEA archives, but I understand that several researchers, including Robert Keohane and Glen Toner, have conducted their work on a similar basis.

I would have assumed far fewer debts over the past decade had I spent my time at the racetrack instead of in the library and the office. A complete list of all the people who have helped me would read like a telephone directory. I would like to begin the acknowledgments by thanking the many librarians, archivists, and scholars who are not specifically mentioned here by name but who contributed so much to the completion of this work.

An early draft of the book served as my doctoral dissertation at the Fletcher School of Law and Diplomacy. I had a demanding yet supportive thesis committee, which included Professors Benjamin J. Cohen, Stephan Haggard, and Robert Paarlberg. I hope they can recognize some of their guidance and wisdom in this work.

The book was written during my tenure as John M. Olin Research Associate in National Security at the Center for International Affairs, Harvard University. I would like to thank the former director of the center, Professor Samuel P. Huntington, for his encouragement of my work in economics and security, and the Olin Foundation for its support. Other center and Harvard colleagues who commented on some or all of the book include Robert Art, Alfred D. Chandler, Robert Keohane, Robert Powell, Scott Sagan, David Spiro, Raymond Vernon, and Nicholas Ziegler.

Ever since beginning the research for the book, I have frequently turned to one source for information and advice. I would like to thank David Painter of Georgetown University for his many comments and suggestions. We met in the Harry S. Truman Library, where he was conducting archival work for his definitive study, *Oil and the American Century*.

My research in Washington was supported by the Historian's Office of the U.S. Department of Energy, where I served as a Visiting Scholar from 1978 to 1979. I am grateful to the former chief historians of the department, Richard Hewlett and Jack Holl, for bringing a relatively young scholar to the office. Roger Anders, Alice Buck, and Terrence Fehner were most helpful in sorting through relevant department archives. I came to the DOE from the Rockefeller Foundation, where I had been a fellow conducting research on solar energy policy. I wish to thank Gary Toenniessen and Ralph Richardson of the foundation, and I recall its late president, Dr. John Knowles, with warm affection.

For making my archival research in Paris possible during the summer of 1983, I must thank Dr. Eric Melby, a gifted U.S. official then posted to the International Energy Agency. My stay in France was financed by a grant from the International Security Studies Program of the Fletcher School of Law and Diplomacy.

I interviewed many former officials for the book, some of whom have preferred to remain anonymous. Among the "open" sources, I am most grateful to the Honorable Nathaniel Samuels, to Mr. Robert Koenig, to Ambassador Milton Katz, and to the late Ambassador Samuel Berger. Nathaniel Samuels took time from his busy schedule on several occasions to discuss with me the European coal crisis. While interviewing these men, I became most appreciative that the United States has had such citizens in government. They provide role models for current and future generations.

My introduction to historical research began when I was an undergraduate at Brown University. My senior thesis adviser, Professor A. Hunter Dupree, was—and has remained—a true mentor. I hope the book provides some reward for all his efforts on my behalf.

Michel de Montaigne wrote something to the effect, "All the world knows me in my book, and my book in me." The members of my family have lived with this project as long as I have, and they have always provided the fuel when I suffered my personal energy crises. I dedicate the book to them.

Cambridge, Mass E.B.K.
March 1989

Contents

Figures and Tables

Figures

Tables

*Maps were prepared by Professor Jacques Lacoste, University of Nancy, France.

The Insecure Alliance

1

Energy Crises and Alliance Politics

"Politics," wrote Michael Taylor, "arises out of disagreement and scarcity."[1] This book, which analyzes Western alliance relations during every major postwar energy crisis, is thus about politics. Fuel shortages have threatened not only the economic well-being of the alliance but also its military preparedness and political cohesion. Although the allies have attempted to formulate a collective response to each energy shortfall, their efforts did not always succeed. The burden of this study is to explain why the alliance response to energy shortages has varied from crisis to crisis and why the West's search for energy security has proved elusive.

The present chapter provides a framework for analyzing alliance politics during periods of energy crisis. In building the framework, I begin with a structural perspective, testing the argument that crisis outcomes have reflected the distribution of energy capabilities within the alliance system. The argument is derived from the "theory of hegemonic stability," a leading paradigm in the literature of international political economy. Hegemonic stability theory, I show, offers a parsimonious account of alliance cooperation and conflict in the energy issue area.

A historical analysis, however, pushes hegemonic stability theory beyond its capabilities. The labels *cooperation* and *conflict* are sparse, and they do not adequately describe the complexity of alliance interactions during crises. My subsequent concern is to understand these interactions, and for this task I borrow from bargaining theory. Bargaining theory is central to the study of crises, "because its constituent elements correspond to what are widely regarded as the most important elements in international behavior—e.g., power, interests, conflict and cooperation. . . ."[2] In highlighting the strategic interests of actors, bargaining theory enriches our understanding of alliance behavior in times of crisis.

I then turn to organizational aspects of alliance energy relations. I discuss why the United States has preferred to manage energy shortages in multilateral organizations, rather than unilaterally or bilaterally. Drawing on the literature of international organizations and international regimes, I argue that multilateral institutions have been utilized primarily because of their effectiveness in reducing uncertainty during crises.

The chapter concludes with a discussion of domestic politics and alliance relations. Domestic actors have often conflicted with the executive branch of the U.S. government over foreign policy, making policy execution costly and, at certain points in history, seemingly unaffordable to elected officials. In many cases, the policy response to crises can only be understood in terms of the preferences of particular domestic actors.

In terms of alliance theory building, I should state that while the book is concerned solely with the energy issue, the case studies of each shortage will hopefully be used heuristically, to "stimulate the imagination toward discerning important general problems and possible theoretical solutions."[3] Since energy sits at the cusp of a broad range of economic and security problems, alliance politics in this area may be suggestive of their relations not only in trade and finance, but in the politico-military sphere as well.

Power and Interests

This book is concerned with the problem of alliance cooperation and conflict during periods of energy crisis. A parsimonious account of cooperation in the international political economy is provided by the theory of hegemonic stability. In its "basic" form, the theory—which is derived from the literature on collective goods—rests on two fundamental propositions: (1) order in world politics is typically created by a single dominant power, and (2) the maintenance of order requires continued hegemony. A third proposition, that *all* states benefit from hegemonic order, remains contested; scholars continue to debate whether hegemonic stability is "benign" or "malign."[4]

Scholars of hegemonic stability cite two exemplars: the Pax Britannica, which endured from the end of the Napoleonic wars until 1914; and the Pax Americana, which existed from 1945 to 1970. During both periods, international economic relations were relatively harmonious, with the liberal ideal of free trade ascendant. The economic systems established by the "hegemons" were also afforded military protection against challengers. Since the hegemon was both the major market for the output of the system and the principal military power, small states were provided with positive incentives to cooperate. Over time, however, hegemons are said to lose their relative power. Other actors assume importance on the political stage. As a result, states begin to "defect" from the hegemon, and systemic fragmentation occurs. Depression and war follow.[5]

Hegemonic stability theory links systemic outcomes to the distribution of "power resources" or capabilities among states. According to Robert Keohane, in addition to military power, hegemony requires a

> preponderance of material resources. Four sets of resources are especially important. Hegemonic powers must have control over raw materials, control over sources of capital, control over markets, and competitive advantages in the production of highly valued goods.

Among the raw materials, Keohane has singled out energy as especially important. By maintaining domestic spare production capacity, and by assisting its multinational corporations in the domination of world energy markets, the United States established order in this issue area after World War II. In so doing, it ensured the energy security of its allies, or access to adequate supplies of fuel at relatively stable prices.[6]

Robert Gilpin has also listed the material requirements for hegemony. He has written that there are "three sources of power in the modern world: nuclear weapons, monetary reserves, and petroleum." Petroleum was cited by virtue of being "the most vital lubricant of industrial economies." Gilpin argues that America's monopoly of nuclear weapons in 1945, and its control over oil and hard currency, gave it the economic and military means to shape the postwar order.[7]

Given the status of energy in the hegemonic stability literature, alliance relations during periods of shortage would appear to provide a fundamental test of the theory (but not a *critical* test, which would require that we take an issue area *not* cited by hegemonic stability theorists).[8] One would posit that alliance cooperation, or mutually beneficial policy coordination, has been a function of whether or not hegemonic resources were available to meet emergency needs during shortfalls. When the United States provided resources, one would predict a cooperative outcome; in the absence of such resources, alliance conflict and a scramble for fuel.

I should make clear why the "power resource" of spare production capacity has been so important to hegemonic stability. Theoretically, leverage in the energy area accrues to those who monopolize supply. So long as the United States maintained spare production capacity, attempts by alliance adversaries to wield the "oil weapon" were undermined; the release of fuel by the United States quickly balanced markets. But hegemonic power has been a two-edged sword for the allies. To be sure, it provided energy security, but it also gave the United States enormous leverage. The exercise of hegemonic power could therefore create intra-alliance tensions, as will be discussed later.

One caveat before continuing. Hegemonic stability theory does not purport to explain crisis behavior per se. Indeed, one could make the case that the very emergence of an energy crisis argues against the existence of a hegemon, since the concept of hegemony suggests systemic order. When I speak of a hegemon I am referring more accurately to an "alliance leader"—namely, the United States. Washington was never able to control the international system after World War II, because another superpower stood in opposition to its values and goals. It did, however, dominate an anti-Soviet alliance. In this study, I test hegemonic stability theory in the context of that subsystem.

A schematic history of the West's postwar energy crises—all of which are discussed in detail in the main body of the book—provides substantial support for hegemonic stability theory. During Europe's postwar coal crisis, the Iranian nationalizaton of British oil assets in 1951, the Suez crisis of 1956, and the Six Day War of 1967, the United States served as Europe's energy supplier of last resort, and alliance cohesion was maintained in the energy issue area each time. In 1973 and 1978–1979, in contrast, when U.S. emergency supplies were absent, the result was alliance discord and panic. The 1980 case, related to the outbreak of the Iran–Iraq war, is somewhat ambiguous; while the allies, notably Japan, initially scrambled for available supplies, some coordination apparently occurred later in the crisis.

On the systemic level, then, hegemonic stability theory offers a parsimonious account of alliance behavior during energy crises. This suggests the importance of U.S. power capabilities—and the willingness to use that power—in promoting alliance cohesion. A coordinated response to fuel shortages has proved elusive in the absence of hegemony.

A detailed historical analysis, however, quickly pushes hegemonic stability theory beyond its limits. Problems arise when we observe closely alliance relations in crisis situations. As stated earlier, the theory would predict cooperative outcomes in the presence of an asymmetric distribution of power capabilities. But during certain crises, alliance disagreements threatened the formulation of a coordinated response, even in the presence of hegemony. At other times, cooperation was relatively easy to achieve. How do we explain these differences in alliance behavior?

The literature on bargaining suggests that it is crucial to understand the *interests* of actors, and the *value* that they place on those interests, in order to explain crisis behavior. *Interest*

> refers to the tangible or intangible things that are at stake for the parties; *value* connotes the kind and amount of utility that are attached to those things. The stakes in a crisis really consist therefore of a set of "valued interests"[9] [italics added].

In the historical section of this book, I will show that energy crises have involved many of the vital political and economic interests of alliance member states. The alliance collective action problem has arisen in large measure from the differing values they placed on those interests. All the allies, for example, have shared a common interest in avoiding a competitive scramble for available fuel supplies, since this would, at a minimum, raise prices. Beyond that, competitive scrambles could destabilize international relations, with consequences that could not be foreseen. But the avoidance of such scrambles has not been the *only* interest of alliance states, or even necessarily the dominant one. States pursue many interests during crises, and intra-alliance conflicts have reflected, in part, the different priorities they attach to their varied interests.

Another way to think about the collective action problem is in terms of hierarchy of interests. The question arises: did the allies have a unity of purpose, a collective hierarchy of interests, during crises, or were their interests simply too divergent? For example, during the Suez crisis of 1956 the allies had shared interests in maintaining oil flows and preventing Soviet military aggression and/or diplomatic gains in the Middle East. But the allies ordered their interests differently. Britain and France sought to maintain control over the Suez Canal Company at any cost, while the primary objective of the United States was to contain the Soviet Union. The clash of dominant interests complicated alliance policy, and a coordinated response followed only after Washington used its oil power coercively, raising the costs of the British and French actions to politically unacceptable levels.

It is curious that the literature on alliance theory makes little mention of this problem of conflicting interests. Indeed, the focus is on the singular common interest that brings states together. Alliances are created when states do not possess sufficient power to maintain their national security on their own. As George Liska has written:

> Alliances are against, and only derivatively for, someone or something. The sense of community may consolidate alliances; it rarely brings them about. . . . Cooperation in alliances is in large part the consequence of conflicts with adversaries and may submerge only temporarily the conflicts among allies.

The crucial tie that binds allies is fear of a common enemy.[10]

This theme of common strategic interests runs throughout the work on alliance creation. Thus, to Hans Morgenthau, the creation of alliances is "not a matter of principle but of expediency."[11] Michael Ward argues that nations form alliances when they share a "commonality of national security interests, typically viewed via a mutually perceived threat."[12] In his usual direct manner, Kenneth Waltz has written that "the common interest" of alliance members "is ordinarily a negative one: fear of other states."[13]

Clearly, the Western alliance was created after World War II owing to a common interest in containing and deterring Soviet military power and subversion. But other common interests have grown out of this larger strategic objective. Relevant to our study, the allies have had a common interest in promoting economic development and social welfare; strong economies provide the allies not only with advanced weaponry but also with political stability and the maintenance of democratic institutions.[14] Economic depression and beggar-thy-neighbor foreign economic policies, in contrast, could lead to political unrest and international disputes. In this context, it should be recalled that Article II of the North Atlantic Treaty (signed in April 1949 creating NATO) pledged that "the Parties . . . will seek to eliminate conflict in their international economic policies and will encourage collaboration between any or all of them."[15]

As I will show in the historical section of the book, energy security, or the maintenance of adequate energy supplies at reasonable cost, has been a common interest of the Western allies since the end of World War II. Energy provides an archetypal example of a natural resource that a modern alliance requires for its military and civilian sectors. Its importance has been stressed not only in the recent hegemonic stability literature but also in the earlier writings of such realists as Hans Morgenthau, Bernard Brodie, N. J. Spykman, and James Schlesinger.[16] That common interest has led the allies to take collective measures in order to ensure their energy security; the allies have also taken various unilateral measures. Alliance policymakers have feared that the failure to preserve their energy security would lead to domestic economic problems, a decline in military preparedness, and, during a severe shortage, an energy scramble that must weaken the alliance relative to its adversary and strain alliance cohesion, perhaps to the breaking point.[17]

HEGEMONIC RESOURCES	COLLECTIVE INTERESTS: YES	COLLECTIVE INTERESTS: NO
YES	**Cooperation** **Postwar Coal Crisis** **Iran, 1951**	**Coercive Coordination** **Suez Crisis** **Six Day War**
NO	**Attempted Coordination** **Iran, 1978–1979** **Iran-Iraq, 1980**	**Conflict** **Arab Oil Embargo**

FIGURE 1–1. Energy Crisis Matrix

Energy security has thus been a "valued interest" of the allies individually and collectively. The problem is that it has been one valued interest among many and that in times of crisis, valued interests may come into conflict. Alliance relations during energy crises, then, have reflected not only the distribution of power capabilities but also the hierarchy of interests of each member state.

Figure 1-1 combines the discussion of power and interests in a highly stylized matrix that generates "predictions" about the outcome of each energy crisis case. I must stress that the matrix is a *heuristic* device, a guide to the history that follows. It is not a replacement for detailed study.

I place along the X-axis "Collective Interests," to indicate whether or not a *unity* of alliance purpose, or a collective hierarchy of interests, has existed during a crisis (I use the term collective interests as shorthand for collective valued interests). The allies, of course, have had many interests in common during energy shortages; the question concerns the priority or relative value they placed on each one. During certain crises, the allies placed similar values on their various interests; during others, priorities differed.

I should note that the X-axis presents only a static view of valued interests. But during crises, the priorities given to various interests may change as actors receive information and inputs from their environment. Further, the axis does not capture the *degree* to which interests have been shared or opposed; alliance differences in 1956, for example, were far greater than those in 1967. The axis simply conceptualizes, in an intuitive way, whether alliance interests were fundamentally harmonious or conflictual.

Along the Y-axis I place "Hegemonic Resources," referring to whether or not the United States provided emergency energy supplies to its allies during shortages. To be sure, energy has not been the only power resource at Washington's disposal; it has employed a variety of political, economic, and military instruments in times of crisis. But the release of fuel has been crucial from the perspective of market balance.

Starting with the northwest cell, the presence of both hegemonic resources and collective interests yields *cooperation*—mutually beneficial policy coordination. Examples are provided by Europe's postwar coal crisis, and the Iranian nationalization of 1951. Alliance policy coordination has been most effective when hegemonic power coincided with a common strategic purpose.

Moving east, I posit that the presence of hegemonic resources and the absence of collective interests leads to *coercive coordination*, as exemplified by the Suez crisis of 1956 and the Six Day War. In these cases, owing to intra-alliance conflict, the United States coerced its allies in order to achieve policy coordination. Washington would not provide fuel to allies whose strategic interests differed from its own.

In the southeast cell, the absence of both hegemonic resources and collective interests yields alliance discord or *conflict*, with the 1973–1974 energy crisis providing an example. The Arab oil embargo was an economic and strategic fiasco for the west.

The southwest cell has collective interests but no hegemonic resources; the Iranian revolution and the 1980 oil shortage caused by the outbreak of the Iran–Iraq war provide the cases. The outcome here is *attempted coordination*. I would posit that so long as the allies have a collective set of valued interests, some efforts at coordination will be made; unfortunately, the effectiveness of these attempts cannot be judged a priori. To be sure, theories of collective choice suggest the severe impediments to cooperation in the absence of a leader, but they do not state that this outcome is impossible to achieve.[18]

I must reiterate that the matrix is not intended to capture forty years of history. It is offered as a road map through the energy crisis cases and as a way of making explicit the key independent variables. The case studies provide a more complicated story and in so doing demonstrate the utility of blending "concepts and realities."[19]

Students of game theory may be disappointed by the fact that I have not described each crisis in terms of such games as "Prisoners' Dilemma," "Stag Hunt," or "Protector." For that matter, Snyder and Diesing have described a class of "alliance games," which focus on intra-alliance behavior in the presence of a common threat.[20] But while it is tempting to reduce international politics to such models, I believe this would interfere with the detailed historical description I seek to provide. Accordingly, I leave that approach to another researcher.

The reader will notice that the cells move in chronological order, beginning in the northwest. In so doing, they suggest a hegemonic life cycle in the energy issue area. In the period of early hegemony, the United States not only held power resources, but it also had a shared set of values with its Western allies; the alliance reflected a combination of power and strategic purpose. It may not be surprising to learn that during both the postwar coal shortage and the Iranian nationalization, the saliency of the Communist threat to energy supplies was high, and that spurred efforts to coordinate energy crisis policies.

As the allies recovered from the war, they began to reassert their independent foreign policy interests. They acted boldly at Suez because, as Glenn Snyder has argued, in a bipolar world they were certain the leader would not "abandon" them to their fate. The United States, however, did not wish to be "entrapped" by its allies' colonial adventures, and in 1956 it acted coercively to raise the costs of their unilateral action.[21] In the short run, this strategy prompted the allies to follow the leader. But over the longer term, it drove them to diversify their energy sources and suppliers, thus contributing to the relative decline of hegemonic power. By the early 1970s, the United States was perceived by the allies to be less important as their supplier of last resort.

With hegemonic decline and conflicting "out-of-area" interests, the allies proved unable to coordinate policy in 1973–1974. For many scholars, the energy crisis marked the end of the Pax Americana, or American century, that had started in 1945. The failure of collective action led to an energy scramble that raised oil prices for all states and, in turn, weakened their economies and military preparedness. It pointed the Soviet Union toward one of the weakest links in the alliance chain, namely Western policy toward the Persian Gulf. During the invasion of Afghanistan, the Soviets were able to lean on this weak link, confident that the allies would prove incapable of a collective response.[22]

But hegemonic decline did not necessarily mean the end of alliance energy coordination. Indeed, I would hypothesize that so long as a set of collective interests endures, efforts to cooperate during crises will still be attempted, as exemplified by alliance relations in late 1980. The possibility for cooperation in a post-hegemonic world exists, but it depends upon the willingness of each ally to shoulder some of the costs associated with their energy security. To put the issue another way, posthegemonic cooperation requires a new formula for energy burden-sharing.

Organizing the Allies

Since the end of World War II the Western allies have attempted to fashion collective responses to energy shortages. Multilateral organizations have been established to deal with energy problems, beginning with the European Coal Organization, which was founded in 1944, and exemplified most recently by the creation of the International Energy Agency in 1974. Conceivably, the United States could have dealt with energy crises unilaterally or bilaterally. Why did it encourage multilateralism in this issue area?

In answering the question, I will focus on the role of information. A concern with information bisects both the literature on crises and on international organization. This suggests the pervasive role that it plays in international politics.

"The essence of crisis," wrote Thomas Schelling, "is its unpredictability."[23] Decision makers must proceed under conditions of uncertainty, and the information they receive "is frequently patchy, of doubtful accuracy, and ambiguous."[24] They do not know the intentions of their adversaries, or even of their friends. Not surprisingly, they often feel that their control over events is tenuous at best.

It is the feeling of uncertainty "that lends to an event its 'crisis atmosphere,' i.e. to feelings of fear, tension and urgency."[25] During an energy crisis, states may lack information regarding the duration and scale of shortages, and about alternative supplies and prices. This uncertainty may lead states to hoard fuel, strike bilateral deals, and take other actions that raise energy prices for all states; the result is a collective bad. An alliance leader would wish to avoid this turn of events. By providing market information to all its allies, the possibility of misperception triggering or exacerbating an energy crisis may be abated.[26]

Couldn't the leader simply provide each ally with equivalent information on a bilateral basis? As Robert Keohane has pointed out in a discussion of international regimes, a multilateral response may lower the "transaction costs" of conducting international relations; multilateral institutions are cost-effective. Further, if information is provided solely on a bilateral basis, each ally might fear that it is not receiving identical data and analysis. This would create an unnecessarily competitive situation.[27]

It is important to stress that the provision of information is not a value-free act. Information is often provided by states in order to *shape* perceptions during a crisis. As Oran Young has written, "The manipulation of perceptions of events or situations in order to shape the bargaining calculations of other parties is generally an important mode of exercising influence."[28] During the Soviet invasion of Afghanistan, for example, President Carter focused on the heightened threat to Persian Gulf oil in order to promote a coordinated alliance response.[29]

The provision of information also provides a positive incentive for states to join in a multilateral organization. In the energy area, it is costly for small states to build the data systems necessary to track market developments, since economies of scale are found in the development and maintenance of data banks. Washington's promise to establish an energy information system in the International Energy Agency compelled some states to join that organization in 1974.

Further, the supply of information may serve a legitimization function. As Raymond Aron has written, legitimate leadership has been important to the United States, and by providing energy information in a multilateral setting, it could espouse the "ideology of equality among allies."[30] Conversely, when it proved incapable of providing accurate information about market developments during crises, the legitimacy of its leadership was cast into doubt.

As I will show in the historical section of the book, the multilateral energy organizations had other functions beyond the provision of information. Specific management tasks included the development of emergency energy allocation schemes and the coordination of member state energy policies. Accordingly, it is possible to match the crisis performance of these organizations against their stated objectives and in so doing learn how effective they have been for managing energy shortages.

At this point I should make explicit why the allies have had an interest in joining the United States in the creation and maintenance of multilateral organizations for energy security. Simply stated, these organizations have helped the allies to realize important national goals. Energy security has been a prevalent concern of the allies since the war's end; multilateral agreements were one route, among others, to achieve the objective of assured access to fuel supplies. To be sure, one could argue that if alliance energy security had been in the strategic interest of the United States, it would have provided this good even if the allies refused to join an organization. But several points can be made in response. First, an ally may have perceived that organizational membership was an efficient means for guaranteeing energy security; in the absence of such membership, costly lags could occur as ad hoc agreements with the leader were struck during crises. Second, as stated earlier, a state could benefit from the access to market information that it gained with organizational membership. Finally, the alliance leader might have coerced its smaller allies, making the costs of refusal to join unreasonably high.

To summarize, from Washington's perspective, multilateral organizations were created because they promoted its legitimate leadership in the energy issue area. By executing their tasks effectively, multilateral institutions also helped the allies realize important national goals. Multilateralism harnessed American power to the advancement of collective strategic purpose.

Domestic Politics and Energy Crises

It has proved impossible for scholars to discuss changes in the international political economy without reference to U.S. domestic politics. By shaping the policies of the hegemon, domestic actors have had a direct impact on international economic outcomes. In studying the regime for international finance, for example, Joanne Gowa has stated that the monetary policies of the United States can only be understood in terms of "the distribution of domestic political power."[31] Similarly, despite his sophisticated systemic analysis of international cooperation, Robert Keohane has concluded that "domestic politics constituted a crucial factor" in shaping U.S. foreign economic policy.[32] And Stephen Krasner, in a study of U.S. foreign oil policy, has argued that "central decision-makers have confronted major difficulties in pursuing national objectives when their policies conflicted with the preferences of particular domestic groups."[33]

The approach to domestic politics adopted in this book corresponds most nearly to that found in the work of Stephen Krasner.[34] Krasner has advocated a "statist" model, which asserts that the President and Secretary of State formulate foreign policies that are viewed in the "national interest" and are consistent across postwar administrations in terms of broad objectives (e.g., containment of the Soviet Union). In the "policy formulation" process, the president is relatively insulated from special-interest groups and is thus able to pursue the interests of the state *qua* state. Taking raw-materials policy as his case study, Krasner claims to have identified the foreign policy goals shared by each postwar administration. In descending order these have included (1) promotion of broad foreign policy objectives, notably containment and deterrence of the Soviet Union; (2) secure supplies of raw materials; and (3) competition and low prices for raw materials. The historical section of this book lends support to his findings.

"However," Krasner has written, "in its pursuit of the national interest, the state may also have to overcome resistance from domestic groups."[35] This is because foreign policies may harm certain societal actors. As an example, the liberalization of oil imports into the United States has had a negative impact on high-cost domestic oil producers, who in turn have organized to fight this policy initiative and to buffer its impact through fiscal policy. Observing this process in the raw-materials area, Krasner distinguished between policy formulation and policy execution. The former may be free of special-interest pressures, but the latter is not. The American political process is characterized by its "fragmentation and dispersion of power and authority."[36] Special-interest groups have multiple channels of access to government decision makers. In raw materials, the state confronts actors "including large corporations and concentrated groups of domestic producers, that possess substantial political resources."[37] These groups may lobby against presidential initiatives, blocking legislation and undermining the executive's foreign policy preferences.

It should be stated, however, that corporate actors may also share a set of specific objectives with the executive branch; interests are not always in opposition. The business–government partnership in Middle East oil development after World War II exemplifies a high degree of policy convergence. At least one scholar has used a "corporatist framework" to describe the close interaction between government and the private sector in the execution of U.S. foreign oil policies.[38]

Krasner's distinction between policy formulation and policy execution is valuable for understanding alliance energy security. Regarding energy policy formulation, I will argue that decision makers have indeed prioritized their goals as Krasner suggests, but that domestic actors often created sufficient political friction to elevate significantly the costs of policy implementation. Leadership by the United States has often been constrained by powerful domestic actors.

And yet I try to avoid using state-level explanations of alliance relations in an ad hoc way. As I will show, it is necessary first to "exhaust" the systemic level of analysis before turning to other levels. Friction among the allies, as at Suez in 1956, created long-term conflicts that contributed to relative hegemonic decline, without any help from specific domestic actors.

Conclusion

The objective of this chapter has been to provide a conceptual framework for the historical study of alliance relations during energy crises. I have argued that a perspective that incorporates insights from hegemonic stability theory, bargaining theory, and the literature on international organizations will prove useful in understanding the case studies that follow. At the same time, the history is complex, and in explaining alliance behavior I will inevitably expand upon and amend the framework presented here.

To recapitulate, I have argued that alliance relations are best understood in terms of the interaction between hegemonic power and valued interests. "Cooperation," or mutually beneficial interaction, required the presence of both hegemonic resources and shared interests. "Coercive coordination," in contrast, occurred when the hegemon used its resources to overcome strategic disputes among the allies. Conflict resulted when both resources and shared interests were absent. In the "posthegemonic" world of no resources but enduring interests, the possibility for alliance cooperation remains, even though it is certainly difficult to achieve. The historical chapters that follow provide a detailed empirical test of this conceptual framework.

2

Europe's Postwar Coal Crisis: 1944–1947

> The restoration of the production and distribution of coal in Europe must be looked upon as the prerequisite to the economic, political, and social rehabilitation of the continent.
>
> U.S. War Department, May 11, 1945[1]

At the end of World War II Europe suffered the most devastating energy crisis in its modern history. The severe coal shortages stymied the continent's recovery and threatened to breed social unrest. Policymakers in Washington feared that many European nations would go Communist if their economies were allowed to collapse. Further, even if democratic regimes could be maintained, it would be impossible under such conditions to advance Washington's objective of building a liberal international economy characterized by free trade, currency convertibility, and an "open door" for foreign direct investment.

In response, the United States played an active role in overcoming the energy shortage and in economic rehabilitation in general. In the coal area, Washington's primary concern was to prevent a scramble for solid fuel that would not only drive up its price but possibly destabilize domestic regimes and international relations. To promote cooperation, a transatlantic institution, the European Coal Organization (ECO), was established in 1944 to allocate scarce solid fuel resources to member states. At the same time, the United States exported domestic coal and mining machinery.

This chapter describes the European coal crisis of 1944–1947. Coal and international politics were inextricably linked during the immediate postwar years, and as a result Washington's response to the energy shortage is best understood in terms of a strategic approach that treats its international economic and security concerns in a unified fashion. Coal was a "high policy" issue in the postwar world, influencing West–West and East–West relations. In this charged economic and political atmosphere, it is a testimony to postwar diplomacy that cooperative solutions to the energy crisis were devised.

The Liberation of Europe

With the invasion of Normandy on June 6, 1944, the allies began to dismantle the political economy of Nazi Europe. It would be only a matter of weeks, however, before the liberators discovered that they had brought no system along as a replacement. The people of Western Europe were economically desperate, requiring supplies of food and coal above all else. Allied troops were unprepared to meet these basic civilian needs.

Planning for liberation in alliance capitals had "not been characterized by clarity of purpose or responsibility."[2] On a general policy level, the United States had pledged to provide reconstruction assistance at the war's end, and it was instrumental in the creation of the United Nations Relief and Rehabilitation Agency (UNRRA).[3] Emergency aid, according to Secretary of State Cordell Hull, was necessary

> not only from a military point of view to the extent that such relief and assistance can preserve order, economic stability and cooperation behind the lines, thus facilitating military operations, but, in the event of a sudden collapse of the enemy, to prevent chaotic economic conditions in the liberated areas. . . .[4]

Hull recognized that a chaotic Europe would doom his ambitious postwar plans. During the war he had fought for the establishment of a world order built on liberal economic principles. He believed that an international economy would promote world peace; the creation of autarkic economic blocs, in contrast, must lead to war.[5]

The achievement of a Hullian order, however, was threatened by economic chaos in the liberated countries and by the competing economic doctrines that thrived in such conditions. As liberation progressed, it was already clear that the Soviets intended to establish a closed sphere of influence in Eastern Europe. Shortly after D-Day, Franklin Roosevelt warned Winston Churchill that Soviet military control of the East would probably "extend into the political and economic fields."[6] According to historian Robert Pollard, American officials feared that the creation of a Soviet economic bloc might "encourage the formation of closed economic blocs in Western Europe and elsewhere." Soviet policies in the East and chaos in the West stood in opposition to Washington's postwar goals, and emergency economic assistance was viewed as the best means for achieving U.S. policy objectives.[7]

The American response to Soviet activity in the East was not just a reaction to postwar events but had historical roots. As stated in an M.I.T. study of postwar aid programs,

> Americans were not unmindful of the danger that communism might spread in the period of postwar disruption. The memory of Communist efforts—successful and unsuccessful—to exploit the opportunities open after the First World War was still alive in the country and in the Congress in 1944–46.[8]

U.S. officials were concerned that Communist doctrine might prove attractive to citizens of a moribund Europe.

Among the economic problems confronting the liberators, few were so challenging as the fuel shortage. World War II "demolished the established framework of energy supply in Europe."[9] The fighting destroyed continental coal mines, oil refineries, and the once-proud transportation network of railroads, highways and inland waterways. The Nazis had used slave labor in the mines, and abandonment of the pits coincided with the German retreat. In Britain the war effort had required overwork of the best mines and a draft of "Bevin boys" (named for Minister of Labor Bevin), males who were ordered for military service to the coal mines. By war's end, Europe's coal mining industry was in ruins, lacking the capital and labor required to restore prewar levels of output.

Prior to the war, coal had met 90 percent of Europe's energy needs, and it was indispensable for transportation, home heating, industry, and the generation of electricity. The exhaustion and destruction of the great European mines had not been foreseen by allied planners (see Figure 2–1). When U.S. Army mining engineer Robert Koenig was asked by the State Department in June 1944 about "the plans for meeting Europe's coal needs after the war," he simply responded, "there are none."[10] In July, Walter Thayer of State's Mission for Economic Affairs (MEA) in London reported to the Pentagon that "considerable confusion exists with regard to the liberated territories of Europe." Thayer feared that if the continent's economic problems were not quickly ameliorated, violence might erupt.[11]

As reports about Europe's coal mines filtered back to London and Washington after D-Day, a disturbing picture of destruction emerged. Mines had been flooded by the retreating Germans, and stockpiles were low. Coal mining machinery was in short supply, as were miners. Europe's forests had been decimated during the war for heating needs, leaving little timber for pitwood. Reconstruction of the coal economy would clearly be a massive undertaking.

On August 22, 1944, MEA officials Sam Berger and Arthur Notman issued a classified report, "The European Coal Problem in the Immediate Postwar Years." This document would prove influential in shaping the postwar political economy. The officials painted a bleak picture of the continent's energy economy. They estimated that Europe would require coal imports of 20 million tons in 1945 and 30 million tons in 1946. Still optimistic about Britain's coal potential—prior to the war it had been the world's largest exporter—they stated that even if the United Kingdom supplied 10 million tons, another 10 million must be found elsewhere. While it was possible that American mines could produce this exportable surplus, U.S. ports lacked loading capacity, and ships were in short supply. Assuming these bottlenecks could be broken, the payments problem loomed large:

> The cost of shipping such quantities of coal across the Atlantic is staggering. Currently, coal is being shipped across the Atlantic at about $22 per

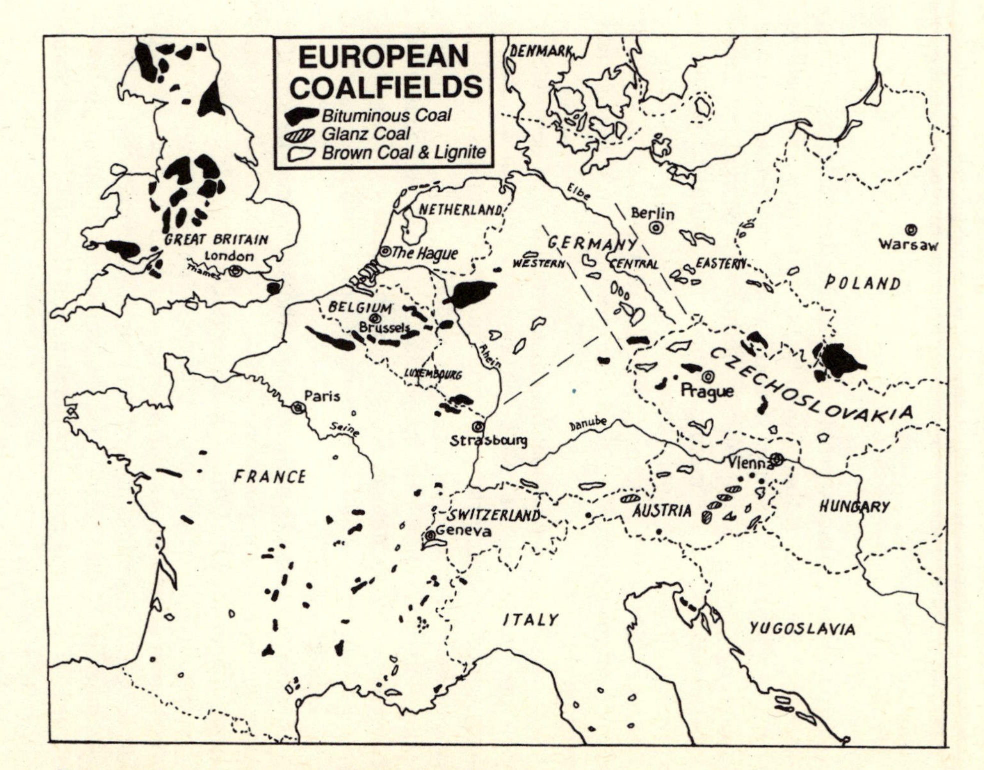

FIGURE 2–1. European Coal Fields, 1945

> ton. Assuming this can be reduced by about one-third to $15 when the war ends, the 1945 imports of 10 to 20 millions from the US would cost from $150 million to $300 million. The 1946 imports . . . would cost from $150 to $375 million. . . . We can draw only one conclusion from the foregoing; namely, that it will be impossible on one ground or another . . . to meet the greater part of the European coal deficit in the first two years after the war. It follows that if the European deficit is to be largely overcome, the coal must come from Europe's own mines or Europe will go without coal.[12]

Despite the enormous problems outlined, Berger and Notman proposed several steps that should be taken by the wartime allies. First, each European nation should submit coal production and stockpiling plans to allied military authorities. Second, all European requests for coal mining equipment should be reviewed. Finally, a "European Coal Organization" should be created "to coordinate the continental program; to examine requests for coal and for mining equipment; to decide priorities; to give technical assistance. . . ."

Why was creation of a coal organization viewed as desirable by some officials in August 1944? The compelling reason was that in the absence of a central allocation authority, nations could be expected to "scramble for available supplies" and "bargain with and outbid each other." The richer nations, especially the neutrals that escaped war damage, would benefit at the expense of the liberated countries, "and some parts of Europe would come to an economic standstill."[13]

In terms of the larger political stage, the reason may be found in America's search for postwar stability. As noted above, foreign policy officials, led by Cordell Hull, were concerned lest economic conditions in Europe become conducive to radicalism and international conflict. If a *sauve qui peut* approach to the coal shortage was adopted, this might imperil not only Allied forces "but also the possibility of re-establishing stable conditions at all in post-war Europe."[14] This was precisely the situation that the United States wished to avoid.

Initial discussions on the ECO were Anglo–American in character, though France, the Soviet Union, and other European nations were soon invited to join the process. While the diplomats argued, the military officers in the field formulated their own policy response. On September 22, 1944, a Solid Fuels Section (labeled "G-4") was created at Supreme Headquarters, Allied Expeditionary Forces (SHAEF). Subsections of G-4 were formed for France, Belgium, the Netherlands, and Germany. This organization was responsible for coal production in the liberated countries and for the allocation of any surpluses produced indigenously or acquired from abroad.[15]

On September 29, Colonel Robert Koenig, the commander of G-4, received a directive from General Eisenhower concerning coal supplies. Eisenhower stated that G-4's goal should be "to assist in the rehabilitation and development of local resources to support the economic effort, to relieve demands on production in the US and UK, and to relieve demands on shipping. . . ." Notably, Eisenhower said that "a European Coal Commission exercising executive authority in the Theatre at this time is not favored." The Supreme Commander wanted his men to run the show, without meddling from politicians or bureaucrats.[16]

The severity of the coal crisis became clearer as liberation proceeded. In December 1944 Koenig estimated that French coal production stood at 40 percent the 1938 level and that it reached 35 percent of the prewar level in the Netherlands and almost 50 percent in Belgium. These shortfalls would have created great hardship even if Europe's traditional surplus producers—Britain, Germany, and Poland—were operating at their 1938 capacity. But this was not the case. The continental coal trade had ceased, while Britain, now producing only 75 percent of its prewar output, was exporting just 100,000 tons of coal per month. American exports, limited by the lack of ships, ran to 200,000 tons per month.[17]

To the men in the field, and to the coal experts laboring in Washington and London, it was clear that Europe's economic salvation lay in the productive might of the Ruhr. Germany had been the continent's principal coal producer, and while mine conditions could not be estimated prior to liberation, it was hoped that many would be rescued intact. Koenig's mining engineers were in the spearhead of the allied attack, reclaiming the mines as quickly as possible.

Yet the issue of German mine reconstruction touched raw nerves in every alliance capital. In the early war years, the allies had failed to take a formal position on Germany's economic role after liberation. The most clearly enunciated proposal was that offered by U.S. Treasury Secretary Henry Morgenthau in late 1944. Morgenthau envisioned a deindustrialized, pastoral Germany that would be broken up into several small states. Execution of the plan would require destruction of German industry and massive population resettlement. Regarding coal, James Riddleberger of the State Department recalled meetings in which Morgenthau "insisted upon writing directives that would have compelled the occupying powers to flood the German mines."[18]

Although most allied governments recognized that it would not be in their interest to reduce Germany to a nation of farmers, crucial differences of opinion remained within and between governments regarding German recovery. The French, and many U.S. State Department officials, advocated a selective recovery in which Ruhr coal output was increased while overall industrial activity remained at a low level. France demanded that German reconstruction take place within a political context that ensured French security. Specifically, the French government of Charles De Gaulle advocated internationalization of the Ruhr, occupation of the Rhineland, and French control of the coal-rich Saar.[19] Secretary of State Byrnes was uncomfortable with the idea of internationalization. As he told French Foreign Minister Bidault in August 1945, "As far as the American government is concerned it does not favor an internationalization which would result in installing Russia, along with Great Britain, France, and the United States, along the Rhine."[20]

The British Government, and the U.S. War Department, saw the concept of a selective German recovery as impractical; one could not rebuild the coal mines in an economic vacuum. Mining equipment and steel were needed as part of the rehabilitation process. Britain, economically weakened by the war, did not wish to prolong the period in which Germany required foreign aid to keep afloat. The War Department, which bore financial responsibility for the U.S. occupation, concurred. Britain and the War Department agreed that an economically reconstructed Germany was in their and Europe's immediate interest.[21]

Given the Ruhr's potential to produce coal for export, allied discussions in London on the creation of a European Coal Organization (ECO) became bound up with the question of German recovery, slowing the negotiation process. But as liberation progressed, the ECO issue grew pressing. The war's end would mean the dismantling of SHAEF and its G-4 Solid Fuels Section, and the western allies agreed that some replacement was urgently needed to allocate available supplies.

By early 1945 the ECO had become a "high policy" issue at the State Department, occupying the Secretary and his closest advisers. The United States strove to de-link the German and ECO questions, so that the coal organization could get off the ground even in the absence of a solution to broader economic recovery issues. But the Russians refused to adopt this approach. Moscow argued that German coal was a form of reparations payment, and thus the reparations issue must be settled before the coal organization could be created. Yet every party knew that reparations negotiations would be a lengthy process, and for the western allies the coal crisis was too pressing to await the outcome of such complex talks.[22]

Coal and the Soviet Union

The United States worked hard to win Soviet participation in the ECO. Acting Secretary of State Joseph Grew cabled London—where the coal talks were being held—in February 1945 to "re-emphasize" State's "dislike of economic regionalism" and its hope that "all continental Europe" would participate in the ECO. Nonetheless, Grew said that the United States would be willing to launch the ECO without Soviet participation.[23]

Why did the United States court the Soviet Union? The answer is found in Roosevelt's international strategy. As John Lewis Gaddis has pointed out, the president was concerned throughout the war "by postwar Soviet power in the case of allied victory . . . containment was much on the minds of Washington officials from 1941 on. . . ."[24] The president favored an approach that Gaddis has labeled "containment by integration." If Stalin was given a major role in the postwar order, Roosevelt reasoned, "cooperation could occur." Roosevelt apparently believed, based on the successful lend–lease experience, that American economic assistance could be used as a carrot for guiding the Soviets, with linkage between assistance and other political issues yielding positive outcomes for both nations. Membership in international organizations was offered in the same spirit.

But Stalin had a different agenda. According to Ambassador Averell Harriman, the Soviet delegate to the ECO talks, Borishenko, was receiving his orders directly from the Kremlin, with no room to maneuver. It was possible that Stalin simply wanted to torpedo the ECO. In a cable to the State Department dated March 21, 1945, the Ambassador suggested that the United States make one last effort to reach agreement with the Soviets but then "proceed with whatever measures we find it in our interest to take . . . independent of the Russians. . . ."[25]

Two weeks later, with the ECO talks still going nowhere, Harriman again cabled Washington. This document demonstrates that "Cold War" tensions were emerging even before the hot war had ended. He said:

> We now have ample proof that the Soviet Government views all matters from the standpoint of their selfish interests. . . . The Communist Party or its associates everywhere are using economic difficulties in areas under our responsibility to promote Soviet concepts and policies and to undermine the influence of the western allies. Unless we and the British now adopt an independent line the people of the areas under our responsibility will suffer and the chance of Soviet domination in Europe will be enhanced. I thus regretfully come to the conclusion that we should be guided as a matter of principle by the policy of taking care of our Western allies and other areas under our responsibility first, allocating to Russia what may be left.[26]

This telegram was of critical importance to officials in the State Department. To Samuel Berger it signaled that "containment by integration" would not work. As Berger said, many policymakers came to believe that the "USSR decided to seize on conditions created by war to create revolution. . . ." The Soviets, according to this view, had no strategic interest in cooperating with the ECO.[27] After the Harriman telegram, such influential policymakers as George Kennan and H. Freeman Matthews (the director of State's European Division) gave up on "trying to negotiate practical agreements with the Soviet Union;" they viewed this cause as "hopeless."[28] On April 11, 1945, the Harriman view was apparently vindicated, as Soviet delegate Borishenko informed the ECO negotiators that his government would not alter its positions regarding the reparations and coal questions.

Western officials such as Harriman viewed Stalin's decision in terms of grand strategy, but a simple economic analysis may have been in order. The fact is, given its military occupation of Eastern Europe, Moscow knew that it could obtain far more coal directly from such producers as Poland and Czechoslovakia than it could by participating in ECO allocation procedures. It probably made no sense to Stalin to share in Western Europe's coal shortage at a time when Soviet consumers were suffering. By grabbing a share of Eastern European coal production, the Soviet Union would be better off than by cooperating with the ECO. Stalin's behavior during the coal crisis certainly did nothing to improve East–West relations, but at least it can be debated whether his actions were an integral part of some master plan.

The European Coal Organization

On May 18, 1945, the European Coal Organization was founded in London by Belgium, Denmark, France, Greece, Luxembourg, the Netherlands, Norway, the United Kingdom, the United States, and Turkey. The Eastern European governments that had been parties to the talks did not join in light of the Soviet decision; however, this was not the end of East–West negotiations on the matter, as will be discussed later.[29]

The ECO was the first international organization created after the European war by the Western allies. Its defined purpose was as follows: first, to promote the supply and equitable distribution of coal and mining equipment; second, to safeguard, as far as possible, the interests of both producers and consumers; and finally, to disseminate information and make recommendations to governments regarding coal supply and distribution. The ECO's primary objective was to allocate available coal supplies to needy member states; it had little input into coal production decisions within liberated Europe and Germany, which were left to national and occupation authorities. The organization was composed of delegates representing each member country, and a permanent secretariat. The chairman of the ECO (J. C. Gridley in 1945 and J. Eaton Griffith in 1946–47) and the delegates comprised the council, which was the decision-making body. In practice, the daily work of the ECO was carried out by various committees, the most important of which were the General Purposes Committee, which discussed all financial issues, and the Allocations Proposal Committee, which made recommendations concerning the country allocations of available coal.[30]

The ECO's recommendations were purely advisory; it had no executive powers whatsoever. "Nevertheless, governments followed its decisions so consistently that its word was virtually final."[31] While the archives do not reveal why the organization lacked explicit authority, domestic politics in the United States and Britain provide part of the answer. President Truman received hundreds of telegrams from citizens in 1945 who were outraged by coal exports to Europe. "I fail to see any justification whatever," one Pennsylvanian complained of the coal shipments, "when our own families suffer because sufficient coal is not available to keep them warm this winter. A coal shortage already exists and I do not believe anything should be done to make a bad situation worse."[32] Similar positions were heard in Britain. Officials in London and Washington probably feared that domestic efforts would be made to block the creation of an organization that had formal powers regarding coal distribution.

The ECO was founded on the principle that during shortages states should not scramble for available fuel supplies, and that coal should be allocated equitably, according to an agreed-upon formula. The rules and decision-making procedures were embodied in formal agreements and were rigorously followed; all decisions were made unanimously. The creation of the ECO reflected in large measure America's postwar interests. The coal crisis was seen as politically divisive, a potential source of European conflict and Communist gains. By harnessing its power in the energy area to its strategic purpose, the United States would not only help its allies to overcome the coal shortage, but advance its broader postwar goals in the process.

The liberated countries welcomed American leadership. At the war's end, the United States was the only nation on earth with spare production capacity in both coal and petroleum. It also produced the machinery and equipment that these industries required, and the ships needed for fuel transport. To most Europeans it was clear that continental recovery would prove impossible in the absence of American assistance.

But why the ECO? Why was a formal organization created in the coal issue area? Seemingly, the United States could have acted unilaterally, drawing up an allocation schedule on its own. Alternatively, it could have negotiated bilateral deals. Instead, a multilateral vehicle was chosen.

The ECO filled several functions. First, while the United States was dominant in the energy area, other nations were important players. Britain was a major coal producer in its own right, and its troops occupied the Ruhr. South Africa was another potential coal exporter. A multilateral organization was viewed as the most efficient means for allocating *all* available coal supplies, not just those from one or two countries. Second, by providing a central information source on coal markets, the organization lifted some of the uncertainty regarding supplies and prices. Third, by allocating coal on a multilateral rather than bilateral basis, the ECO helped to create an atmosphere of trust, and to legitimate U.S. leadership. Finally, the ECO was used as a forum for issue linkage, in which the United States attempted to link resolution of the coal crisis with other trade and financial policies. The ECO also had strong practical appeal because of the history of Europe's coal industry. Before the war, Britain and the continent had been linked by an intricate coal network composed of numerous importers and exporters. The war destroyed this network, and it could not be rebuilt by unilateral or bilateral measures. In short, only a multilateral approach made sense.

Technically, the ECO worked as follows. Each month (later quarterly), the organization was informed by national authorities, and by the occupation forces in Germany, of any exportable surpluses. These supplies were in turn allocated to members that requested them on the basis of a formula that took into account such variables as prewar consumption, war damage, and indigenous resources. Amounts allocated on this basis were called "pool A." If supplementary amounts were available for a given period, this "pool B" was shared among those countries that could use the fuel to bolster internal coal production. All allocation decisions were made on the basis of unanimity.[33]

It must be stressed that the American coal exported to ECO members was *sold*, not given, and had to be paid for in U.S. dollars. Coal from Germany and other European mines was paid for with a mix of currencies, but primarily dollars. These dollar payments soon proved burdensome to a dollar-short Europe; as will be seen, this "dollar shortage," caused by European imports of American goods and services, ultimately led Washington to propose a comprehensive European Recovery Program, known as the Marshall Plan.

Increasing German Coal Production

When the ECO was founded in mid-May 1945, the war in Europe had just ended. The battle for Germany was hard fought, leaving the once great power in shambles. The invaders quickly recognized that German industrial recovery would take years.

For Robert Koenig, who had crossed the European continent as head of SHAEF's G-4 section, no picture of destruction was more grim than that of the Ruhr. Allied bombers had pounded the region to rubble. Over one third of the housing stock was in ruins, and transportation was completely immobilized. The manufacture of machinery had ceased. The slave laborers had quit the mines, leaving a skeletal work force. When the Ruhr mines were seized in April 1945, production stood at 30,000 tons per day, down from a daily wartime average of 400,000 tons.[34]

Surveying Europe's coal crisis at the war's end, Koenig made an eloquent plea for a cooperative international response:

> The kaleidoscopic nature of this vast and complicated problem places upon the many agencies, whether they be private enterprises, governmental or military, engaged in its solution both in liberated countries and occupied areas, a great responsibility. These agencies will not be dealing solely with physical things, but will necessarily be influencing the social and political status of millions of people, only some of whom are directly engaged in the coal industry. This is a challenging problem.

It should engage the minds and will of the best men that industry and government can produce.[35]

Koenig argued that Europe's coal shortages could only be relieved by indigenous production, especially in the Ruhr. Britain could offer no more than 100,000 tons per month, and American exports must be limited by the expense of carriage and the lack of ships. Now that victory over Germany was complete, decisions had to be made regarding its economic rehabilitation. The work of ECO and other economic relief organizations would be of marginal value if Germany's productive capacity were not mobilized for European recovery.

At liberation, Germany was cut into occupation zones by the allies; Britain took control of the Ruhr and exercised authority over the zonal economy. While an allied council in Berlin was supposed to decide Germany's economic fate, this system of governance proved untenable. Not only East–West differences, but clashes among the Western allies, hindered the policy formulation process.

Despite the previous pleas of Berger, Notman, and Koenig, the most influential "high policy" argument for a rebuilt Germany was presented by an Anglo–American team of coal experts, C. J. Potter and Lord Hyndley, following an official tour of the major European coal mines in June 1945. The report was presented to President Truman and Prime Minister Churchill on June 19, and it had an immediate impact on Western policy. The team urged:

that it be recognized by the highest authorities in the United States and Great Britain that, unless immediate and drastic steps are taken, there will occur in Northwest Europe and the Mediterranean next winter a coal famine of such severity as to destroy all semblance of law and order, and thus delay any chance of reasonable stability.[36]

In order to meet the shortfall, at least 30 million tons of coal would have to be made available over the next year above and beyond Western Europe's projected output. Of that amount, it appeared that the United States, Britain, and South Africa could contribute only 5 million tons. This meant that Germany, and Poland if possible, must contribute 25 million tons. Potter and Hyndley urged the allied leaders to take "quite extraordinary steps to stimulate the production in, and movement of coal from, western Germany." Zonal commanders should provide food, clothing, and shelter to the miners. But Ruhr coal must be mined to meet *European* as opposed to *German* needs. Only the minimum amount of solid fuel necessary to meet industrial production levels should be allocated to German industry and citizens; accordingly, allied forces must be prepared "to deal with any unrest which may result. . . ."

The Potter–Hyndley report formed a compelling link between coal supplies and European stability. Five days after the report reached Washington, President Truman cabled General De Gaulle and Prime Minister Churchill with his response. He said:

> The coal famine which threatens Europe this coming winter has impressed me with the great urgency of directing our military authorities in Germany to exert every effort to increase German coal production and to furnish for export the whole quanitity over and above minimum German needs.

The president recommended that at least 25 million tons of German coal be made available for export by April 1946 and urged that each commander in the western zones be given firm directives to maximize output.[37]

The Truman telegram was not welcomed by America's military governor in Germany, General Lucius Clay. He told the War Department's John McCloy,

> I am not sure that anything approaching these (coal production) figures can be met. I am sorry to see coal treated separately from the economy of Northwest Europe as a whole. . . . The successful large-scale mining of coal means some restoration of the German economy and some industrial activity in Germany to support coal mining.[38]

At the Potsdam conference in July 1945, the first elements of a tripartite (Soviet–U.S.–U.K.) policy for controlling the German economy were supposedly established. The wartime allies called for the stimulation of mining and immediate measures to repair transportation, utilities and infrastructure. But despite the veneer of agreement, basic differences concerning German economic recovery went unresolved. The French, who were not invited to

Potsdam, felt no obligation to adopt the conference resolutions in their sector or the Saar. For their part, the Russians continued to focus on reparations, dismantling much of the industry in their zone for shipment to the Soviet Union. The debate over the "level of industry" appropriate for postwar Germany would prove a divisive issue for some time to come.[39]

As the winter of 1945–1946 approached, it was clear that continental Europe would suffer from severe coal shortages. An interested observer of the seasonal change was United Mine Workers' president John L. Lewis, who found in the coal problem a crisis of opportunity. In the autumn of 1945, Lewis called a strike, causing U.S. coal export to slow to a trickle. Under Secretary of State Will Clayton, assessing the impact of the strike on U.S. foreign policy, asked the Secretary of Labor to intercede. A prolonged strike, Clayton said, "may result in conditions of such distress and economic dislocation as would threaten the maintenance of stable governments and law and order on the continent of Europe."[40]

The ability of labor to undermine American foreign policy suggests, in the words of political scientist Stephen Krasner, the "paradox of external strength and internal weakness." Policy execution in the United States has often been blocked by important domestic actors, who hold the "national interest" hostage to narrow demands. This theme will reemerge often in our study of alliance energy security. It highlights the fragility of U.S. foreign policy and the difficulty of harnessing power to serve strategic purpose.[41]

American officials went on an offensive to end the strike. Returning from an autumn trip to Europe, Interior Secretary Harold Ickes told Clayton, "I would not like to have on my conscience any part of the responsibility for the tragic failure of the country to produce all the coal which it can. . . ." Ickes, the last New Dealer still in the cabinet, worked hard with the UMW to reach an agreement. The strike ended on October 22.[42]

ECO officials were relieved by the news, coming just as winter loomed. Then, two months later, the organization was presented with a Christmas gift as Czechoslovakia joined the organization. The Czech foreign minister, Masaryk, pledged to make some coal and coke available for export. While the Czechs were not major coal producers, this East–West thaw indicated that Masaryk retained some latitude in foreign policy making and that sufficient production capacity existed in Eastern Europe to more than satisfy immediate Soviet coal demands.[43]

In February 1946, after months of negotiations, Poland also agreed to ECO membership. Yet Poland joined under special conditions that proved troubling to some member states. First, Poland was permitted to maintain its existing coal export agreements with the Soviet Union, shipping westward only the surplus amounts that existed above these contractual levels. Second, a Polish official was granted the position of ECO vice-chairman. Finally, Poland was allowed to reserve a portion of its export coal for bilateral trade with the neutral countries. While Polish membership would have been welcome at almost any cost, some ECO delegates wondered if these conditions didn't change the character of the organization. After all, if Poland engaged in bilateral dealing, what would prevent other nations from following suit? Uncontrolled bilateral exchanges for coal would destroy the ECO's very *raison d'être*.[44]

There is a poignant element to this swan song of East–West relations. In the summer of 1946, as Poland was negotiating with the United States for some mining machinery, a Polish diplomat pleaded, "Don't leave us behind the Iron Curtain." The ECO represented a final glimmer of hope for the London Poles who had been allied to the West during World War II. Soon all remnants of Western influence in the government would be eradicated, and Poland would become a Soviet satellite.[45]

Ironically, the addition of two Eastern European countries to the ECO took place at a time when East–West tensions were growing in Germany. In January 1946, France called for the internationalization of the Ruhr by the Western allies out of "fears that Germany may become part of the Russian orbit and that the production of the Ruhr might be used by a hostile Germany and Russia against France."[46] The French claimed that Moscow was not interested in Western European recovery; on the contrary, it preferred chaos and depression, which would favor national Communist parties. Many State Department officials were sympathetic with this viewpoint, arguing that the French had a "good case." In a classified document one top regional expert said, "the lines between the east and the west are daily becoming more sharply defined." But the United States was not prepared to renege on the Potsdam objective of centralized allied control over the German economy.[47]

At the same time, the French were expressing their dissatisfaction with Britain's policy in the Ruhr. Sensitive to French complaints, Secretary of State James Byrnes dispatched two of his economists, John Kenneth Galbraith and Walt Rostow, to Germany to report on coal production. On April 30 the team confirmed that Truman's "25 million tons" directive would not be fulfilled and that Ruhr exports were not "sufficient for French recovery." The French argument that Britain was mismanaging the mines appeared justified. The economists stated that food rations were low, machinery was unavailable, and housing conditions were abominable. Worst of all, Nazi-era managers were still in charge of the pits, creating terrible morale problems for the workers. One year after liberation, production stood at 900,000 tons per week, less than half the wartime level. Galbraith and Rostow recommended that greater pressure be placed on British authorities to offer the miners adequate food and clothing, and such luxuries as cigarettes, in order to bolster coal production.[48]

State Department officials Clayton and Dean Acheson placed the coal problem in a larger political context. In a cable to Byrnes, who was attending the Council of Foreign Minsters meeting in Paris, they wrote

> Our view is that US is vitally interested that immediate steps be taken to increase coal production in Ruhr. European coal production now limiting factor on entire European recovery. . . . US economic policy for Europe . . . requires that increasing amounts of German coal be made available to help recovery in France and other Western European countries. . . . US position is clearly sympathetic to French request for increased exports, keeping in mind necessity for gradual revival of German industry at a lower rate than that of liberated areas.[49]

This balancing of French security requirements with German economic recovery would prove one of the enduring themes of the early postwar period.

Coal and France

The Acheson/Clayton telegram signaled an important victory for French officials. Since early 1946, Jean Monnet had been developing a Modernization and Re-Equipment Plan (the Monnet Plan) which provided the political–economic foundation for French postwar recovery. Monnet and his colleagues were drawing a detailed blueprint for the economy, with specific production targets for such "key industries" as iron, steel, and coal. The objective of the plan was to provide not just for France's economic security but for its military security as well. As historian Alan Milward has pointed out, the Monnet Plan "had important international as well as domestic intentions. One aim of modernization was to make the French economy more internationally competitive in the future, particularly against German competition." This meant that the French, on the one hand, must have assured access to Ruhr coal; on the other, the reconstruction of German heavy industry must be curbed. So long as Ruhr industry was impotent, the French had no reason to fear German aggression or economic competition. Paris consistently pursued the twin goals of obtaining the maximum amount of Ruhr coal while limiting the expansion of German industry throughout the period of reconstruction.[50]

But coal shortages threatened the entire plan. On May 18, 1946, Monnet cabled his old friend Will Clayton to report that the lack of fuel created a "critical situation for France." The country was attempting to recover from war using 20 percent less coal than the 70 million tons consumed in 1938. Imports of German coal were 300,000 tons per month, half the prewar level, while British coal imports were only 20 percent of prewar. American coal, which was entering France at a rate of 400,000 tons monthly, posed a heavy burden on the treasury; the transatlantic crossing cost $20 a ton. Monnet said that greater quantities of less expensive Ruhr coal were required to avert an economic catastrophe. "The coal problem," he told Clayton, "must be regarded as the most important economic problem facing Europe."[51]

Monnet argued that execution of his "plan" also required continued French occupation of the Saar. Under French management, the region was producing 51 percent of its prewar levels. Monnet contrasted this figure with Ruhr production, where the British had only achieved 35 percent of prewar output. State Department executives found Monnet's arguments compelling, particularly in 1946, when they feared that the French Communist Pary might win or capture a substantial share in the upcoming national elections.[52] Economist Charles Kindleberger recalled officials at State warning that "France will go communist if the demands of the French for coal . . . are not met."[53] Europe's coal shortage was not viewed as a simple economic problem but as a complex strategic one.

The French position regarding Ruhr coal was given further support at an international coal conference sponsored by the ECO in Paris in May 1946. The delegates, representing twenty-one states and four international organizations, agreed that German coal exports should be maximized while German consumption was held in check. Four resolutions were passed at the conference:

> (1) that unless exceptional measures are taken at once, the shortage of solid fuel in Europe over the ensuing twelve months is likely to be such as to cause widespread unemployment and seriously to retard postwar recovery; (2) that this catastrophe can be avoided only if effective steps are taken to ensure adequate food for miners . . . to provide special incentives to attract . . . labour to the mines . . . to give special priority to the manufacture and distribution of mining equipment . . . to ensure the availability of adequate facilities to lift and transport every tons of coal mined; (3) that the above measures . . . are essential if dislocation of almost all industries . . . is to be avoided; (4) that the application of the above measures cannot in itself result in fulfillment of the minimum needs of Europe unless accompanied by rigid control of consumption of coal in Germany and maximum export of coal from that country.[54]

This widespread support for the French position solidified the State Department's position that Paris must obtain more German coal. For State, U.S. strategic interests in Europe were best served by an economically robust France. This meant that Ruhr coal must be exported, even if German industry suffered as a result.

To the Marshall Plan

The spring of 1946 found major issues besides Ruhr recovery on the ECO agenda. One controversy centered on a Belgian–Polish proposal to permit ECO members to engage in barter exchange for coal. To nations that had suffered relatively little war damage, this idea held obvious appeal; consumer and industrial goods could be traded for energy. But for countries like France and Italy which produced few surplus goods for export, the idea was threatening. The proposal was soon vetoed by the United States. As Averell Harriman, now in London, explained in a cable to the State Department:

> US objective should be to utilize ECO machinery to develop a trade pattern that will approach as far as possible ultimate aim of free multilateralism. In view of basic importance of coal . . . we believe US can convert situation described above into an opportunity for developing a major European trade pattern consistent with US general trade aims. Practices developed now will have a strong tendency to continue beyond emergency.[55]

Harriman's message reveals the broader U.S. policy goals that were embedded in functional organizations like the ECO. Despite Europe's economic problems, a "Hullian" approach to the international economy maintained its doctrinaire hold on Washington. To European governments, which were struggling with inflation, shortages, and black markets, talk of "free multilateralism" must have appeared hollow.

Secretary of State Byrnes offered Harriman strong support for the position taken in ECO meetings. Barter agreements, he said, would defeat "the main purpose of the ECO." He observed that bilateral trade would benefit the neutrals that had "suffered no war damage." Further, Byrnes reminded the delegates that the United States "is probably in best position to supply world needs and thus in best position to make barter deals, but has rejected many attractive opportunities to do so."[56] The implicit threat was clear: if the Europeans chose not to play by the ECO's established rules, the entire game could change to their detriment.

The seriousness of America's quest for a liberal international economy was also demonstrated by its decision in late 1946 to place all coal purchases in private channels. Under the second War Powers Act, the Treasury Department had been authorized to act as a middleman, purchasing coal for Europe from private suppliers and in turn receiving payment from the ECO member states that received the coal. Anticipating the expiration of the act in 1947, the Truman administration decided to hasten the shift from government to private hands. This shift was disturbing to cash-strapped Europe. The United States had been a lenient supplier, basically providing ECO members with short-term loans in lieu of immediate payment. But given Europe's credit standing after the war, private-sector firms would demand "cash on the barrel." ECO members were left wondering whether the United States had really come to terms with the continent's economic crisis.[57]

Indeed, faced with the prospect of a U.S. retreat from the financing of coal purchases, ECO chairman J. C. Gridley sought to move the drama to the United Nations. In November he wrote Secretary General Trygve Lie:

> The largest single factor holding back the progress of reconstruction in Europe . . . is the acute general shortage of solid fuel. . . . In placing before you the broad facts of the all important coal position there is of course no thought in my mind of . . . promoting financial loans from the United States to any country. On the other hand, I feel sure you would wish that a matter of such cardinal importance . . . be fully faced and be the subject of separate and urgent action by the United Nations Secretariat.[58]

While the United Nations would not take up the financial aspects of Europe's coal shortage, it did promote numerous activities in the energy area, including sponsorship of conferences, publication of statistical series, and the conduct of detailed studies on regional energy balances.

As winter descended upon Europe in late 1946, economic conditions steadily deteriorated. Now Mother Nature sabotaged the recovery effort, slamming Britain and the continent with the worst storms in fifty years. Blizzards halted transportation and production in industries and mines. Electricity blackouts were widespread as utility poles were knocked to the ground. In response, the British government imposed severe energy conservation measures, and it even began to negotiate for American coal through the ECO. During the winter, British unemployment stood at two million workers; Anthony Eden declared that Britain was "confronted with the gravest industrial crisis that has faced us at any time in the last 20 years." It was during this awful winter that American officials began to recognize the need for a new approach to European recovery.[59]

By early 1947 the coal crisis, the growing Cold War marked *inter alia* by the Greek civil war, and Europe's deepening financial problems led American policymakers to consider a comprehensive assistance program for Europe. On March 12, President Truman requested from Congress $250 million for Greece and $150 million for Turkey. In his dramatic "Truman Doctrine" speech, the president declared that "economic and financial aid . . . is essential to economic stability and orderly political processes. . . . The seeds of totalitarian regimes are nurtured by misery and want."[60]

At the same time, Truman's new Secretary of State, General George Marshall, ordered that an interagency task force produce a study of Europe's requirements for economic and military aid. A State–War–Navy Coordinating Committee was subsequently formed, and it indicated that a multibillion-dollar effort would be required. The committee pointed to Europe's need for food and coal and to the material shortages that prevented continental industries from reaching their productive capacity.[61]

Parallel studies were conducted by the State Department's new Policy Planning Staff (PPS), under the direction of veteran diplomat George Kennan. In its first report, the PPS argued that America "should select some particular bottleneck or bottlenecks in the economic pattern of Western Europe and institute immediate action . . . in the breaking of these bottlenecks." Kennan suggested that coal production in the Ruhr should be the first object of attack.[62]

The PPS devoted its next paper entirely to the European coal problem. The staff argued that Britain and America must work jointly to restore German and Polish mines and set as an objective the restoration of output to prewar levels. Such a program, the PPS estimated, would reduce by $650 million European expenditure for American coal. The staff felt that it would take at least "two or three years" to achieve this goal.[63]

On May 8, 1947, in a speech to the Delta Council in Cleveland, Mississippi, Under Secretary of State Dean Acheson brought the message about Europe to America's heartland. He detailed the enormous destruction wrought by war, focusing on the shortages of food and fuel. Europe, he pleaded, needed America's help.

Will Clayton added his voice to those focusing on the coal crisis. Returning from Europe at the end of May, Clayton said that "Europe must again become self-sufficient in coal." But he disagreed with those who gave a prominent role to Britain in the recovery program, saying that "the US must take over management of Ruhr coal production."[64]

On June 5, 1947, Secretary of State George Marshall offered American aid for European recovery in a speech at the Harvard commencement. Marshall told the graduates that Europe was trapped in a "vicious circle" that could only be broken with outside assistance. He said that food, fuel, and dollar shortages were plunging the continent toward "economic, social, and political deterioration of a grave character." If Europe worked together to formulate a *joint* and comprehensive recovery plan, the United States would be prepared to help.[65]

With the Marshall Plan the United States finally tied its economic and security objectives together into a strategic package. It thrust the United States into a leadership role, even though many Americans still preferred isolation. Indeed, having pledged economic assistance, Marshall would now have to sell his plan to Congress and the American people.

The "Marshall Plan summer" that followed the Harvard address saw feverish activity on both sides of the Atlantic. In July, the sixteen European nations that accepted the Secretary's aid offer met in Paris to create the Committee for European Economic Cooperation (CEEC), which would be responsible for producing the joint recovery document. Simultaneously, several interdepartmental and Cabinet committees met in Washington; Congress also formed a committee, under Christian Herter of Massachusetts, to assess the domestic economic impact of a major American aid program.[66]

The Soviet Union, it should be stressed, attended the initial Paris meetings but ultimately rejected participation in the Marshall Plan. The Secretary had won a calculated gamble in that his offer had not ruled out Soviet participation, even though it was unwanted. Ultimately, the Soviets refused American aid because of strategic considerations. "The Soviets," argues Robert Pollard, "who equated economic control with political mastery of Eastern Europe, probably regarded the American goal of an open, free-trading, economically integrated Europe as a thinly disguised challenge to their sphere of influence."[67]

While the various Marshall Plan groups labored during the summer of 1947, an Anglo–American conference on Ruhr coal production was held in Washington. The U.S. and British zones in Germany had been fused as "Bizonia" since January 1947, and now the coal problem was indeed a joint one. The conference delegates agreed to a twelve-point program which had as its objective Ruhr output of 300,000 tons per day by year's end, up from the current level of 240,000 tons. The allies pledged to provide more food to the miners, better housing, and modern machinery. Further, the Germans themselves would be given greater responsibility for mine management, with the allies serving mainly as bankers and advisers.[68]

Conclusion

While the ECO continued its allocation function throughout 1947, the many changes of that year spelled the organization's end, and in December it disbanded by the unanimous decision of the member states. Western Europe was now joined in the CEEC, with a coal subcommittee, composed mainly of ECO delegates, responsible for determining how much U.S. aid would be required to rebuild the coal industry. A new organization with rather vague functions, the United Nation's Economic Commission for Europe (ECE), had been founded in Geneva, and ECE Secretary Gunnar Myrdal suggested that ECO's statistical work be shifted to his organization. The work of the ECO was thus splintered among the new institutions that dotted Europe's political landscape.

The ECO, which allocated over 100 million tons of coal between 1945 and 1947, was a remarkably effective organization. It did its job efficiently, and it created a cooperative atmosphere in the midst of crisis. At the time the ECO folded, journal articles read like obituaries mourning the death of a prominent citizen.[69]

Several factors contributed to the ECO's success. The extreme severity of the postwar coal shortage, and the interdependence of Europe's coal network, favored a multilateral solution. No state doubted that the war had caused the coal crisis, and every country suffered. Unilateral responses would have surely failed given the complex mining, transportation, and distribution system devoted to coal supply.

But collective action may have failed without a leader, and the United States played a crucial role in the foundation and maintenance of the ECO. The United States charged for its coal, but at least it made fuel available at a time when domestic demands were great. Further, American diplomatic action stifled the few internal disagreements that threatened the organization.

Yet another element was the quality of the ECO's professional staff. The organization was manned by the continent's most eminent coal experts. It would have been impossible to criticize the organization on grounds of incompetence. Generally, ECO officials were mining engineers, not politicians, and they knew their business.

Finally, ECO had a focused objective. Its primary task was to allocate coal to member states in the fairest possible manner. While this involved tricky negotiations over the allocation formula, it prevented delegates from getting sidetracked by tangential issues.

The ECO was the first formal transatlantic organization created to deal with an energy crisis. It established the principle that states should not scramble for available fuel during shortages and that supplies should be allocated equitably, according to an agreed-upon formula. This concept would remain at the heart of alliance energy-sharing arrangements.

The history of Europe's postwar energy shortage reveals that coal and international politics were inextricably linked. The coal problem posed the specter not just of economic depression but of political destabilization. American policymakers believed that an energy "scramble" would favor the Soviet Union and its Communist Party associates in Europe. It would doom the creation of a liberal international economy.

Of course, the ECO was unable to overcome many of the obstacles to greater European coal production; as Table 2-1 reveals, coal production in 1947 was way below prewar averages in almost every European country. Complex East–West and West–West disputes prevented the execution of clear policy guidelines in such

TABLE 2–1. European Coal Production, 1938–1947
(in OOO metric tons)

Country	*1938*	*1945*	*1946*	*1947*
Belgium	29,585	15,833	22,784	24,300
France	47,562	35,017	49,298	50,500
Holland	13,659	5,225	2,958	3,200
West Germany (Ruhr)	138,400	35,800	55,300	74,800
United Kingdom	231,000	186,000	193,000	199,000

Source: Committee for European Economic Cooperation, *Technical Reports* (Washington: Department of State, 1947).

regions as the Saar, the Ruhr, and Poland. The United States, despite its economic and military power, found its leadership blocked at the German border by the Soviet Union. With the Marshall Plan, the United States became committed to building—and fueling—a strong Western alliance.

3

Energy and the Marshall Plan

> Europe in the next ten years may shift from a coal to an oil economy and therefore whoever sits on the valve of Middle East oil may control the destiny of Europe.
>
> SENATOR OWEN BREWSTER, May 2, 1947[1]

The Marshall Plan had a major impact on, among other things, Europe's energy economy. Over the period 1948–1951, the United States provided the participating countries with $1 billion in aid for oil purchases and $400 million for coal and machinery, out of a total economic recovery package of $13 billion. The Europeans also used their own funds to make substantial investments in the energy industry.

More important, decisions taken during the Marshall Plan years resulted in dramatic changes in Europe's energy economy. Of greatest long-term significance, Europe began to shift from reliance on indigenous coal to dependence on Middle East petroleum. The United States formulated a policy of "hemispheric" oil self-sufficiency, which was designed to meet its international economic and security concerns. At the same time, Washington discouraged Western Europe from becoming overly dependent on Polish coal; decision makers believed that the Kremlin could easily translate dependence on solid-fuel exports into political leverage.

The Marshall Plan period also saw the emergence of a "united Europe," with the establishment of the European Coal and Steel

Community (ECSC). The creation of this organization marked a watershed not only for the European states but for transatlantic relations as well. As will be seen, the ECSC assumed many of the coal allocation functions that were once the province of the European Coal Organization.

This chapter opens with a discussion of coal and the Marshall Plan, examining sectoral developments in the context of the Cold War. The following section analyzes Marshall Plan oil policies and the policy decisions taken to encourage European consumption of Middle East petroleum. Overall, energy for European recovery posed more than a narrow economic problem for Western alliance policymakers. It posed a strategic problem of great complexity, involving West–West and East–West relations. The decisions taken during the Marshall Plan years cannot be understood apart from the broader political issues that shaped international relations.

Coal and the Marshall Plan

Energy policymaking for the European Recovery Program (ERP—the Marshall Plan's official name) commenced soon after the famous speech at Harvard. The assistance program was based on the belief that a "dollar shortage" impeded European recovery. This dollar shortage was a condition in which Europe demanded more in imports from the United States than it could earn through exports; coal imports were a major cause of this problem. The objective of the recovery program was to increase production in European industry, agriculture, and mines and thus lessen demand for American goods and services over time. ERP officials estimated that exceptional levels of aid would have to be provided for four years, to 1951. The continent's balance of payments problem, in turn, would gradually be rectified through greater exports and dollar earnings.[2]

In Paris, the newly created Committee of European Economic Cooperation (CEEC) was divided into numerous specialized subcommittees that represented each major economic sector; examples include the coal, oil, and electricity committees. These working groups, manned with European industry experts, were charged with the task of indicating the amount of U.S. aid that the particular sector would require during the four-year recovery program. Another task was to project the production levels the sector would achieve. The CEEC's overall goal was to develop a plan that would permit Europe to surpass prewar industrial output levels by 1951.

While these planning activities occurred in Paris during the summer of 1947, parallel efforts were ongoing in Washington. An interdepartmental European Recovery Program Committee (ERPC), involving principally the Departments of State, Commerce, Treasury, and Defense, had been established shortly after Marshall's announcement, and its job was to reach an independent judgment regarding Europe's need for aid in various economic sectors. As in Paris, the American working groups were composed of some of the government's most competent officials, including such economists as Charles Kindleberger, Louis Lister, and Willard Thorp.

These U.S. officials perceived Europe's economic crisis as a security issue. As one diplomatic historian has written, "U.S. policymakers feared that without massive dollar aid the European economy would deteriorate so much that most of continental Europe would fall into the hands of the Communists."[3] Marshall Plan strategy during the summer of 1947 combined a sense of economic and political urgency.

Given all the attention paid to the coal problem after World War II, one irony of the Marshall Plan summer was the prospect that coal would only play a limited role in European recovery in the foreseeable future. State Department economist Louis Lister, in an ERPC background study, projected that even if Western Europe achieved production of 461 million tons of coal in 1949 (1947 production totalled 350 million), it would still require 46 million tons of imported coal that year at a cost of $850 million. A shortage of indigenous solid fuel appeared to be an economic fixture, creating a serious bottleneck in the recovery effort.[4]

From the outset, Marshall Plan coal policy was framed within a broad strategic context. While the stated goal of coal policy was "recovery of European self-sufficiency," this was to be achieved in a manner that minimized dependence on Soviet bloc producers, such as Poland. Indeed, Cold War considerations were decisive in shaping Marshall Plan energy policies.[5]

U.S. policy accordingly focused on reconstruction of Western

Europe's leading producers, Britain and Germany. The program for Britain fell into two major phases, over the short and long terms. The short-term program was designed to maximize existing mining methods by introduction of new machinery and concentration of miners in the most productive pits. The long-term objective was to develop open-pit mining. The total program was expected to cost over $1 billion, part of which would be subsidized by Marshall Plan funds.[6]

In Germany, the recovery program called for Ruhr output of 440,000 tons per day by 1953, an increase in five years of over 30 percent. This $480 million program would involve "rehabilitation of present mines, new shaft sinkings, major increases in power-generating capacity, repair of bomb damage, and extensive underground development." In addition, a variety of incentives were to be offered miners to enhance morale and increase productivity.[7]

These ambitious proposals would still leave Europe dependent on imported coal, principally from the United States, by the end of the Marshall Plan, with the dollar expenditure such shipments entailed. Another source that had the potential to help overcome Western Europe's shortfall was Poland. But Poland's role in European recovery had posed a dilemma for the United States ever since 1945. The problems were stated succinctly by Paul Porter, acting chief of the Mission for Economic Affairs, in a 1946 memo to Will Clayton. Poland's importance to Europe had been greatly enhanced, he said, by its position of being a coal exporter. In the absence of American coal shipments to Europe, it would become the continent's leading supplier and certainly an "indispensable source of supply to many coal-deficit countries of Europe." But it was unclear whether or not the United States should bolster Polish coal production by exporting needed machinery or encourage Western European imports of Polish coal. As Porter said:

> A great deal of uncertainty as to the future arises inevitably from Poland's relations with Russia. A Russian-dominated Government of Poland might well be tempted to use Poland's strategic coal position as a powerful economic weapon for political purposes. The most vulnerable countries would be France, Italy, Austria, Switzerland and the Scandinavian countries. Three of these, France, Italy and Finland, have strong Communist parties. There has already been a tendency on

> the part of Communist ministers in France and Italy to exploit coal agreements with Poland for partisan aims.[8]

But a policy of limiting dependence on Polish coal might lead to a coal-short Europe that impeded economic recovery, which in turn would play directly into Communist hands; this was the policy dilemma.

Economist Louis Lister had projected during the summer of 1947 that if machinery was made available to the Silesian mines of Poland, perhaps 35 million tons of export coal per year could be produced. But Lister, like Porter, also noted the security risks. Sizable imports of Polish coal would give "to a country within the USSR bloc tremendous economic power in Europe." The economist could only ask his superiors, "What is US policy with regard to that country?"[9]

An answer was provided by State Department counselor Charles "Chip" Bohlen. Bohlen responded that while the United States would not aid Eastern Europe directly, given the Soviet bloc's rejection of Marshall Plan aid, it was prepared to grant some World Bank and Export–Import Bank loans to Poland.[10] These loans, however, would not provide enough capital to bolster Polish coal production to the maximum levels thought feasible by Lister; in any case, it was not in America's interest to make full use of Poland and promote Western European dependence on that supplier.

Yet within the State Department, uneasiness remained over U.S. policies toward Poland. As late as 1949 the belief remained that U.S. aid could be exploited "to draw Poland closer to the west."[11] The reader will recall the remark of a Polish coal official cited in the previous chapter who pleaded to an American counterpart, "Don't leave us behind the iron curtain." This ambivalence toward Poland, and Eastern Europe in general, characterized U.S. policy throughout the early postwar years.[12]

Since Poland was not regarded as a reliable supplier of solid fuel, and given the impediments to mine recovery in Germany and Britain, it appeared unlikely that Europe would achieve its prewar consumption levels during the Marshall Plan years. A draft report prepared for the CEEC's coal group during the summer of 1947 estimated that Europe's energy requirements must be met by "the expansion of hydroelectric capacity and the increased use of petroleum products. . . ."[13] State Department economist Charles Kindleberger pointed out that even if Europe were able to mine coal at prewar levels, much of it would sit at the pithead for want of railway cars and barges.[14]

Still, the coal committee in Paris remained optimistic about Europe's coal production potential. During the four-year recovery program, the CEEC members hoped to boost their output from 414 million to 511 million tons. This would be accomplished by the use of modern machinery, incentives to attract more labor to coal mining, and an increase in the supply of pitwood and steel supports for rebuilding the mines. It was estimated that $607 million of coal mining machinery from the United States and other suppliers must be imported to achieve these production goals.[15]

The Europeans ventured to Washington in October 1947 to discuss their recovery plans. Technical working parties of American and CEEC officials compared notes on economic projections for each sector. In general, the tone of the meetings was one of Euro-optimism offset by U.S. pessimism. This was especially clear in the energy talks.

In the case of the Paris report on coal, U.S. delegate John Havener described it as "inadequate and inaccurate." Havener, who had worked on coal policy throughout the war, believed that Europe had made unrealistic projections regarding the availability of transportation and mining equipment and thus doubted that the coal production targets could be achieved. He further chastised the CEEC for failing to consider the application of new technology, especially strip mining.[16]

An even worse transgression in Havener's view was the fact that the study represented a conglomeration of *country* reports, as opposed to a joint *European* plan. According to the CEEC, U.S. coal mining machinery would not be allocated according to mine productivity but by country, with each participating state obtaining a share of the equipment. This alone meant that Europe would not maximize indigenous production. Havener urged the CEEC to create a "super-intergovernment agency" to oversee the coal sector.[17]

The American's response to the CEEC suggests the degree to which the "macropolitical" concept of European union had trickled down to the microeconomic level. State Department counselor Chip Bohlen had made clear in a 1947 memo that a major objective of the ERP was an "eventual European customs union." The Europeans, he said, "must take concerted efforts to foster European recovery as a whole. . . ." Yet few U.S. officials believed that European federalism would emerge spontaneously as some Platonic ideal. It had to be built sector by sector.[18]

The American government's objective of creating a "united Europe" transcended every economic sector. In some cases the policy appeared quite doctrinaire. During the CEEC talks, for example, the U.S. expressed disappointment that little thought had been given to interconnecting Europe's electricity grid. G. R. Peterson of Britain pointed out that large-scale interconnection programs per se were of dubious economic value. Trade in electrical energy would occur in appropriate circumstances, but not simply as a matter of policy.[19]

German Coal

Throughout the life of the European Recovery Program, coal policy for the Ruhr remained a focal point of policymakers. "The real aim of the Marshall Plan," wrote an American official, "is . . . the appropriation of German raw materials and products for the reconstruction of Europe."[20] On the coal production side, German managers were again assuming control of the mines' daily operations. These activities were supervised by the UK/US Control Group, which was established on November 19, 1947, ten months after the creation of the Bizone. The ECA supplied 60 percent of the capital invested in the Ruhr mines during the Marshall Plan years, and additional funds went to miners' housing and support services. Overall, Germany received $1.4 billion in ERP funds, making that country the fourth largest recipient, behind France, Britain, and Italy. As a result of this investment, Ruhr coal production climbed to 110 million tons in 1951, short of the 1938 level of 138 million but way above 1946 output of only 55 million tons.[21]

On the export side, an International Authority for the Ruhr (IAR), composed of Britain, France, the United States, and the Benelux countries, was founded in December 1948. The purpose of the IAR was to control "distribution of coal, coke and steel of the Ruhr in order that on the one hand industrial concentration in that area shall not become an instrument of aggression, and on the other will be able to make its contribution to all countries participating in a European cooperative economic program." The creation of the IAR symbolized the shift in international politics that had occurred since the war's end. Whereas the ECO had been formed to distribute *all* available coal to member states, and its membership included Poland and Czechoslovakia, coal allocation had now become a Western alliance issue. Further, it signified European acceptance of the idea that an economically renascent Germany was indispensable to the continent's recovery.

Yet German recovery had to take place on terms acceptable to its wartime victims. This meant that German coal production must be made available to France and the Low Countries, even at the expense of German consumption. Further, it meant that the output of Ruhr industry would have to be supervised by an alliance body. With the IAR the French obtained much of what they wanted from Germany: assured access to Ruhr goods at stable prices, and oversight of the region's industry.

France's victory, however, would prove short-lived. Planning was already under way in Western capitals for the creation of a reconstituted German state. The French had to accept that the Ruhr must form part of this new state and that doubts would be cast on the legitimacy of their claims to the Saar. In September 1949, the Federal Republic of Germany was created; one month later the Soviets countered with the formation of the German Democratic Republic. The division of Germany made official a Cold War that had chilled East–West relations since the end of World War II. Paris would now have to find a way to live with the German phoenix.[22]

The birth of West Germany posed a problem for France. It had fought throughout the recovery period for access to Ruhr coal. Now, its economic and security policies were threatened. An independent Germany could fashion Ruhr industry as it wished, since the strategic importance of Germany to the Western alliance meant that Bonn could now make strong arguments for bolstering the domestic economy, even at the expense of lower coal and steel exports for France and other European states. Clearly, Bonn would not accept the low domestic consumption levels that had been forced upon it by the postwar occupation authorities.

The Schuman Plan

The specter of a revitalized Germany once again dominating Europe haunted Franco-German relations in late 1949 and early 1950. French officials scrambled to devise a new strategy that would preserve the state's economic and military security. As it had done so often in the past, Paris turned to Jean Monnet for a solution.

On May 9, 1950, after the completion of preparatory studies by Monnet, French Foreign Minister Robert Schuman proposed that the coal and steel industries of France and Germany, and any other European nation that wished to participate, be placed under a supranational authority in order to create a single market for these goods and to forward the cause of European union. The objectives of this "Schuman Plan" were threefold: first, to ensure continued French access to critical German goods; second, to begin the reconciliation of Franco-German differences in order to prevent future conflict; third, "on the international plane . . . not only to strengthen Western Europe in face of the Russian threat but also . . . to strengthen it vis-à-vis its indispensable but overpowering American ally."[23]

The Schuman Plan had an immediate financial purpose as well. In 1950, the end of Marshall Plan aid was in sight. Monnet believed that, given America's insistence on European integration, a concrete step might merit continued economic assistance.[24]

The plan won the support of Germany, Italy, and the Low Countries, but Britain declined to participate. The British refusal was a blow to French plans. The reasons for rejection included Britain's interest in maintaining an independent role in the Western alliance as head of the "sterling bloc" and as "special partner" of the United States, the economic views of the Labour government toward the capitalist-oriented coal and steel community, and Britain's refusal to surrender its sovereignty to a supranational institution.[25] Still, the United States backed the Schuman Plan, and Washington played an important if quiet role during the initial negotiations, promising to provide financial support to the community should it be approved. As William Diebold has commented, "the United States was lending its weight and diplomacy to help European governments complete an undertaking they had begun. . . ."[26] Washington's position was that the ECSC represented an important step toward the political and economic unity that lasting postwar recovery demanded.

The ECSC assumed many of the allocation functions that had originated with the ECO before being passed to the International Authority for the Ruhr. During coal crises, the coal and steel community was granted broad powers. "In periods of shortage the High Authority [the executive branch of the ECSC] may . . . establish by unanimous vote consumption priorities and an allocation of the coal and steel resources of the Community. . . ."[27] The principles of the European Coal Organization had endured, as member states of the new coal community pledged to avoid a scramble for solid-fuel resources in the event of a crisis.[28]

The creation of the ECSC denoted an important shift in European coal supply patterns. While the continent still relied on the United States for marginal coal shipments, it was no longer so dependent on American solid fuel as it had been during the early postwar years. American coal exports, which had risen above 20 million tons during the late 1940s, fell to 7.5 million in 1953. Increased production, the result of Marshall Plan aid and increased investment, offered hope that the worst of the coal crisis was over. From 1951, Europe would become increasingly responsible for its own coal affairs. This was just what Monnet wanted. Even though the ECSC was financed in large measure by dollar-denominated bonds sold in New York, the creation of the community indicated that Europe was beginning to wean itself away from the United States.[29]

Still, the formation of the ECSC supported broad American strategic goals. Of greatest importance, many U.S. policymakers believed that a European union was necessary not only for the continent's recovery but also in order to make the Western allies a bulwark against Communism. Accordingly, the United States encouraged economic integration by providing financial aid and assistance. The community also held the promise of preventing European member states in future from engaging in a divisive competition for fuel supplies should coal shortages again become severe. These benefits far outweighed any costs that might be associated with an increasingly independent Europe.

The Schuman Plan launched a Franco–German *rapprochement,* which formed the core of the new Europe. But it did not resolve all outstanding issues between the two nations. Indeed, observers of Schuman's gesture noted the irony of a coal-and-steel plan that did not include a settlement of the Saar conflict. Nonetheless, despite the compelling arguments in favor of a quick return of the Saar to Germany, it proved difficult for the politically divided French government to take decisive action.

As discussed in the previous chapter, the Saar was of considerable economic importance to postwar France, given its coal mines and proximity to French industry. Initially, De Gaulle had pressed for internationalization of the entire Rhine valley to ensure French security, but this proved unacceptable to the United States. As a result, France dissociated its Saar, Ruhr, and Rhineland objectives, a decision made easier by French military occupation of the Saar. But the French Army presence meant that Paris felt no compulsion to reach a political agreement. This was a completely different situation from that pertaining in the Ruhr, where France had to rely on diplomacy to meet its strategic objectives.

As Jacques Freymond has remarked, "Properly speaking, there was no Saar conflict until the end of 1947 because of the predominance of France. But from 1948 the situation changed. Opposition arose which became more pronounced as Germany recovered."[30] The Saar conflict also became enmeshed in Cold War politics, as Germany assumed a crucial position in the Western alliance. A vital issue for the United States was to balance German and French demands in order to maintain governments that were friendly to Washington. Thus, the French could no longer expect as sympathetic an audience in the United States as they had earlier received.

Officially, the United States continued to support the separation of the Saar from Germany and its economic integration with France. At the same time, the autonomy of the Saarlanders (or *saarois,* as the French optimistically called them) had to be assured. Beneath the surface, however, the United States was altering its position during the Marshall Plan years. As Freymond has noted, "By 1950 the Federal Republic had become not only a possible and useful ally but a necessary one who must be treated with consideration. . . . The memory of war was fading as a new threat grew."[31] The shift in Washington's approach was manifest not only in a disinclination to make public statements in support of French claims but in press reports about the Saar conflict as well. Bonn's demand that the postwar status quo be reconsidered was now viewed with sympathy in the United States.

Increasingly, France saw its position erode, and by the early 1950s it accepted the right of the Saarlanders to decide their own political fate in a referendum. Ironically, domestic politics in France may have prevented an earlier resolution of the conflict, one that could have turned out in France's favor. But by the time a plebiscite was actually held in 1955, the German economy was the most dynamic in continental Europe, making it a more attractive partner than France. Ultimately, the Saar would return to Germany, the result of a combination of forces that included the Cold War, German economic recovery, and French domestic politics.

Summary

During the Marshall Plan years it proved impossible to restore European coal production to prewar levels and to make the continent energy self-sufficient once again. In part, this was because coal issues were treated in a strategic rather than purely economic context, creating difficult tradeoffs for Marshall Plan policymakers. Such issues as aid for Polish coal production, the future status of the Ruhr and the Saar, and the creation of the European Coal and Steel Community were among those that had to be faced during the period 1948–1951. In some cases, the policy choice resulted in less coal for Europe than might have been economically feasible.

Purely domestic reasons also existed for the failure to produce more coal. In 1950, Britain, for example, produced less than 220 million tons of coal, against a 1938 level of 231 million. Despite nationalization of the industry in 1947 by the Labour government, management–worker relations remained poor, and productivity lagged. Many small and inefficient mines were kept in service, as members of Parliament protected their constituents. Indeed, the problems of the coal mining industry had roots going back to World War I, and a four-year recovery program, no matter how enlightened, could not overcome them all.[32]

Yet the United States could also point to achievements in the coal issue area. German coal production rose dramatically, from a 1948 level of 87 million tons to 1951 output of 119 million. Most

TABLE 3-1. European Coal Production, 1948–1950
(OOO metric tons)

Country	*1938*	*1948*	*1949*	*1950*
Belgium	29,585	26,690	27,850	27,300
France	47,562	43,290	51,200	50,840
Holland	13,659	11,030	11,700	12,250
United Kingdom	231,000	212,740	218,600	219,760
West Germany	138,400	87,030	103,240	110,760

Sources: Organization for European Economic Cooperation, *Coal and European Economic Expansion* (Paris: OEEC, 1952); Committee for European Economic Cooperation, *Technical Reports* (Washington, D.C.: U.S. Department of State, 1947).

important, with the exception of a temporary bulge caused by the Korean War mobilization of 1951, the postwar trend of increasing U.S. coal exports to Europe was curbed, with the concomitant promise of future dollar savings.[33]

Still, as Table 3-1 illustrates, extensive aid to the European coal sector would not bring production up to prewar levels during the Marshall Plan years. The failure of coal prompted Europe's transition to Middle East oil. This shift, which would accelerate up to 1970, created new challenges for alliance energy security. As will be seen in the following sections, many of these challenges were already revealed during the Marshall Plan years.

Oil in Postwar Europe

I have argued that U.S. policymakers regarded the provision of adequate energy supplies to Western Europe as essential for economic recovery and political stability. The European energy problem, however, was complicated by the fact that prewar supply patterns could not simply be reconstructed. Whereas before the war Poland had been a major coal exporter, Cold War politics prevented it from reassuming this role. In oil, the western hemisphere had been Europe's principal supplier, but postwar concerns about the inadequacy of American reserves led officials in Washington to seek new sources of supply.

During the Marshall Plan years, Europe began the energy transition from reliance on indigenous coal to dependence on Middle East oil. Although petroleum's share of the overall energy economy remained small in 1951, climbing from under 10 percent in 1938 to approximately 15 percent by the ERP's conclusion, an infrastructure of refineries and industrial boilers was emplaced that permitted increasing use of liquid fuel throughout the 1950s. At the same time, U.S. policies encouraged the development of Middle East concessions by its national companies.

The shift away from indigenous coal supplies raised a host of security issues that worried American officials. Middle East oil flows could be easily interrupted by any number of regional and external actors. The Soviet Union had posed a threat to regional stability since the war's end, placing constant political pressure on Turkey and Iran. Local military forces could cut pipelines and harass tankers; as Defense Secretary James Forrestal remarked, foreign oil could be cut off "by any third-class navy or air force. . . ."[34] Given the political situation in Palestine, which served as the terminus of several Middle East oil pipelines and as the region's refinery center, it was likely that oil exports would be interrupted.

Yet European coal could hardly be regarded as a secure resource. The coal mines of Western Europe were populated by union laborers with radical and occasionally violent tendencies; France had been wracked by vicious miner disputes many times since the war's end. One of the continent's leading coal producers, Poland, was a Soviet satellite. Furthermore, continental production still had not reached prewar levels.

As one response to the coal crisis of 1944–1947, Western Europe had already increased its oil consumption. In July 1946, for example, British Minister of Fuel and Power Emmanuel Shinwell announced a policy "to encourage the conversion of some industry and transport from consumption of coal to the utilization of petroleum products." This policy included four elements: first, government assistance in procuring oil-burning equipment; second, tax incentives for investment in such equipment; third, subsidization of oil prices; and finally, elimination of customs duties on oil imports. It was hoped that these measures would bolster fuel oil consumption by 2 million tons annually.[35] Upon repealing the duty on oil, Chancellor of the Exchequer Dalton lamented, "It is not a joyful thing, but it is a national necessity to import more oil."[36]

Similar policies were enacted on the continent, and the State Department observed this nascent shift to oil with concern. Acting secretary Will Clayton cabled his ambassadors in July 1946 with the following warning:

> Department has observed recent trend in a number of countries of conversion from wood and coal to fuel oil. This trend apparently inspired by lack or high cost of coal. . . . If plans for conversion to fuel oil continue there is a strong likelihood that sufficient fuel oil will not be available on long-term basis to satisfy requirements. Caution in conversion is therefore dictated.[37]

American anxiety was rooted in numerous postwar analyses that foresaw a serious oil shortage. During the war years, U.S. oil production had soared to meet alliance needs, but price controls and a lack of equipment led to the depletion of reserves. In 1937 the United States was a net oil exporter; in 1947, for the first time in its modern history, the country was a net importer, and dependence on foreign oil was projected to climb. Consumption in the United States totaled 271 million tons in 1947, against production of 253 million.[38]

World demand was also rising. Prewar Europe relied on oil for only 8 percent of its energy requirements; given the coal shortage, this figure would inevitably increase. But Europe had relied on the western hemisphere for nearly 80 percent of its oil during the 1930s. In 1946, the United States and Caribbean still met 75 percent of Western Europe's requirements. With America now in a deficit position, pressure was building in Congress to halt petroleum exports. Economist Willard Thorp of the State Department reported to George Kennan, "There is considerable anxiety in Congress over the large U.S. exports of oil to Europe. . . . U.S. domestic supplies of oil are currently strained."[39]

Emerging postwar oil developments seemed to support the petroleum policies initially developed by the State Department during World War II. In April 1944, the department issued a paper entitled "Foreign Petroleum Policy of the United States."[40] The background to this paper was provided by consultant Everett DeGolyer's mission to the Middle East at the beginning of the year. During his survey of Persian Gulf resources, DeGolyer observed, "The center of gravity of world oil production is shifting from the Gulf–Caribbean area to the Middle East . . . and is likely to continue to shift until it is firmly established in that area."[41] DeGolyer drew two conclusions from this historic shift: first, that western hemisphere oil must be conserved; second, that eastern hemisphere oil must be developed with dispatch. This two-pronged approach would form the core of America's postwar oil policy.

Building on DeGolyer's findings, the State Department's petroleum paper suggested that foreign oil policy rest on five principles: (1) the "equal access" clause of the Atlantic Charter; (2) encouragement of American enterprise in the search for foreign oil; (3) implementation of a broad policy of conservation in the western hemisphere "in the interest of hemispheric security, in order to assure the adequacy for military and civilian requirements of strategically available reserves;" (4) American access to sufficient quantities of eastern hemisphere petroleum to permit the United States to "take appropriate part in a system of collective security;" and (5) the economic development of oil-producing states.

Postwar events did not contradict this hemispheric approach to oil policy. If Europe's oil needs were to be met, this must be done from the Persian Gulf. At the same time, if America's energy security was to be preserved, exports must stop, and western hemisphere resources must be conserved.

Thus, as officials in Paris and Washington pondered Europe's energy future during the Marshall Plan summer of 1947, Middle East oil emerged as an attractive resource. Although production in the region was limited prior to and during the war, it expanded rapidly after V-E day, as Table 3–2 demonstrates. Iran's output, for example, doubled between 1937 and 1947. Saudi Arabia's prewar production was negligible, while in 1947 it totaled nearly 90 million barrels. Kuwait came on stream after the war, with 1947 production of 16 million barrels. These numbers are all the more impressive when one considers the severe drilling equipment limitations under which the oil companies labored at the time.[42]

Before the war, the British-dominated firms of Anglo–Iranian and Royal Dutch/Shell had a near monopoly in the Middle East.

TABLE 3-2. Middle East Oil Production, 1945–1950
(Thousands of Barrels)

Country	*1945*	*1946*	*1947*	*1948*	*1949*	*1950*
Iran	130,526	146,819	154,998	190,384	204,712	242,475
Iraq	35,112	35,665	35,834	26,115	30,957	49,726
Kuwait	NA	5,931	16,225	46,500	90,000	125,722
Saudi Arabia	21,311	59,944	89,852	142,853	174,008	199,547

Source: DeGolyer and MacNaughton, *Twentieth Century Petroleum Statistics* (Dallas: DeGolyer and MacNaughton, 1984).

This would soon change. In December 1946, Standard Oil of New Jersey and Mobil joined the original Saudi concessionaires, Socal and Texaco, to create Aramco. The resources of these four companies were devoted to the rapid exploitation of their Persian Gulf prize. Gulf Oil had already entered into a joint venture with Anglo–Iranian in Kuwait. Overall, American firms were soon rivaling the British in Middle East production, and by 1950 their shares would be almost identical. As will be discussed later, the State Department played a useful role in helping the firms, interceding on their behalf to win valuable Commerce Department export licenses for steel and oil equipment.

Despite the rapid growth in crude oil production, several hurdles prevented Europe from rapid expansion of petroleum use. Price controls in the United States were lifted on June 30, 1946, and by 1947 oil prices had doubled, from $1.32 per barrel to $2.68. Since U.S. oil prices determined the world price, petroleum loaded in the Middle East quickly followed suit. Given Europe's dollar shortage, this rapid price inflation made the cost of oil imports prohibitive.[43]

Still another hurdle was Europe's lack of refining capacity to process foreign crude. On the eve of World War II, for example, Jersey Standard had ten refineries in Europe that churned out 60,000 barrels per day of product. "On V-E day," according to Jersey's house organ, *The Lamp*, "they could not refine a single barrel of crude oil." A reporter for *The Lamp* recalled the sight that greeted American oilmen at liberation:

> Most of the refineries were twisted, burned and gutted ruins when the bombing ended in 1945. Bulk plants and terminals had been destroyed, their storage tanks collapsed and fire blackened. Tankers had been sunk at sea. Tank barges had gone down at their loading docks. Railroad tank cars had been blasted to bits or had vanished into thin air, to be found months and even years later, wandering over half the railroads of Europe.

With new materials and equipment in short supply, oilmen went to work with the one item that was plentiful: scrap. Europe's refineries were literally "salvaged from their own remains."[44]

Yet a further challenge to increased reliance on Middle East oil was posed by the Soviet threat. During the immediate postwar years, British and American policymakers grew increasingly suspicious of Soviet designs in the Persian Gulf. Stalin, for example, had signed a treaty with Great Britain and Iran in 1942 in which the pledge was made to withdraw all occupying armies from Iran within six months of the war's end. In 1946, British and American troops departed, but Soviet forces remained in place.

Indeed, in late 1945 the Soviets sponsored an uprising by the Azerbaijanian population in northern Iran, to which the Kremlin responded by moving Red Army tanks into the region, ostensibly to bring order. Secretary of State Byrnes demanded the immediate withdrawal of Soviet forces and warned of a firm American response if this was not done. In March, Moscow and Tehran began negotiations on the terms of the Soviet departure; one demand placed on the table by Stalin was that a joint Soviet–Iranian oil company be formed. Ultimately, the Soviets backed down, and the Red Army left Iran in May. As historian Walter LaFeber has written of the incident, "The Soviets . . . suffered a major diplomatic defeat." Nonetheless, Moscow had demonstrated its willingness to use military force to advance its strategic aims in the region.[45]

Oil and the Marshall Plan

The economic and political uncertainties surrounding Middle East oil made U.S. policymakers wary of encouraging European reliance on this source of supply. As in the case of coal, the Washington talks on petroleum during the autumn of 1947 pitted Euro-optimism against American pessimism. In its Paris report, the CEEC set aggressive targets for petroleum use. Consumption was projected to reach 72 million tons in 1951, double the 1938 level. The majority of this oil was to be processed in new European refineries. Prior to the war, Western Europe imported 62 percent of its petroleum in the form of refined products. In an effort to save dollars, the CEEC sought to build an indigenous refinery sector, so that by 1951 only 30 percent of Europe's oil imports would be crude petroleum. Still, the costs of the oil program would be great. The CEEC allocated $584 million for crude imports from "dollar" (i.e., American) companies and $575 million from "non-dollar" (primarily British firms selling oil for sterling) sources. A further $1.6 billion would be required for dollar refined products, and $1.9 billion for sterling products.[46]

Before analyzing the CEEC program in detail, the sterling–dollar distinction must be highlighted, since it was of more than academic interest during the Marshall Plan years. Simply stated, dollar oil refers to the crude oil and refined products sold by U.S.-based firms for American currency. Nondollar oil refers to the petroleum sold by foreign firms, mainly Anglo–Iranian and Royal Dutch/Shell, for foreign currency, primarily sterling. At this time, sterling was not freely convertible into dollars or gold. In light of the severe dollar shortage, consumers outside the United States preferred sterling to dollar oil. With many countries encouraging the use of sterling oil, American multinationals feared a severe diminution of their market share and a consequent threat to their concessions. This sterling–dollar oil problem became a major Anglo-American dispute during the European Recovery Program, as will be detailed later in this chapter.

The Washington talks on petroleum were among the most divisive held between the European and American recovery teams during the autumn of 1947. Representing the United States were oil experts Walter Levy and John Loftus. Levy, an emigre from Germany, was an experienced oil consultant who had served with the OSS during the war, tracking German oil shipments with alarming accuracy. Loftus was the longtime head of the State Department's Petroleum Division. Facing this team was British oilman Angus Beckett, who had enjoyed a notable career in industry and government.

The Americans felt that world oil supplies were too tight—and would remain so over the recovery period—to meet Europe's projected demand. The United States was unwilling to export oil in substantial quantities, given its own deficit position, and sought to conserve western hemisphere resources for security reasons. Middle East oil, in contrast, would someday be plentiful, but probably not before the mid-1950s. Loftus and Levy pointed out that exploitation of foreign oil would be slow owing to domestic demand for steel and oil equipment, and it was unlikely that these goods would be exported to the Persian Gulf in substantial quantities.[47]

Going further, the Americans worried about "the wholesale shifting from coal to oil" that they perceived in Europe's energy plans. They questioned the "wisdom of a continent which is seeking to achieve equilibrium in its balance of payments from converting indiscriminately from a material available indigenously to one which requires foreign exchange expenditures." While Loftus and Levy recognized Europe's need for oil, they stated that the CEEC program must be scaled back considerably, especially in those areas such as industrial boilers where coal and oil competed directly.

Beckett countered by insisting that petroleum was absolutely vital to Europe's industrial production goals. He avowed that "the oil, coal and other energy programs . . . have been correlated within the CEEC . . . so as to avoid duplication of energy requirements from various sources. . . ." He also reminded the Americans that petroleum was severely rationed throughout Europe. Gasoline use for pleasure driving, for instance, represented only 12 percent of gasoline consumption.[48]

It must be emphasized that it was the Europeans rather than the Americans who pushed for greater use of oil during the Marshall Plan. American reservations were based on economic and strategic factors. From an economic perspective, it appeared that the oil shortages would endure for another two or three years, with high prices draining Europe of precious dollar supplies. Politically, dependence on Middle East oil, at a time of regional instability, appeared unwise. Oil would certainly be needed for lubrication and transportation, but a rapid increase in consumption should not be encouraged. Overall, the Washington talks deflated Europe's recovery projections.

If the Europeans hoped that, with the creation of the Economic Cooperation Administration (ECA) in Washington to supervise the Marshall Plan, new officials would supervise petroleum policy, they received a shock; Walter Levy was appointed head of ECA's petroleum division. Before discussing oil policy further, some background on the ECA must be provided.

In 1948, Congress created an Economic Cooperation Administration to carry out the European Recovery Program. A businessman, Studebaker president Paul Hoffman, was appointed ECA Administrator. Congress wanted to assure American taxpayers that they would get their dollar's worth out of the European Recovery Program; foreign aid would not be dispensed by the bureaucrats in Foggy Bottom.

Congress also gave itself the right to debate and renew the "Foreign Assistance Act" each year over the four-year recovery period and thus to change the budget and the act's amendments. This gave the House and Senate substantial control over the scope and direction of Marshall Plan aid.[49] As will be discussed later in this chapter, Congressmen exercised their power during the life of the plan to advance constituent interests.

In practice, the Marshall Plan worked as follows. Each quarter the Organization for European Economic Cooperation (OEEC—successor to the temporary CEEC) placed procurement orders for goods and services. These invoices were analyzed by the ECA and then approved, rejected, or amended. If accepted, the ECA paid a private supplier to ship the goods. At the same time, the receiving government had to place "counterpart funds," the dollar equivalent amount of U.S. aid, in its central bank. These funds were to be used by the government for investment purposes and could only be disbursed upon approval of the local ECA representative.[50]

The ECA was composed of two parallel organizations—the headquarters in Washington (ECA/W), and the Office of the Special Representative (OSR) in Paris. Each had staff to analyze Europe's various economic sectors. The first OSR chief was the ubiquitous Averell Harriman.

This structure did not remove the State Department from the recovery program. Indeed, Paul Hoffman and his colleagues worked closely with State Department officials. It must be emphasized that the ECA's primary goal was administration of Marshall Plan funds; larger policy issues (e.g., European integration) were still handled by the White House and the State Department.

At the beginning of Levy's tenure as petroleum division chief, a severe oil shortage still gripped the United States and the world. The Foreign Assistance Act, which authorized aid for European recovery, took into account the American shortfall by requiring that the "procurement of petroleum and petroleum products . . . shall, to the maximum extent practicable, be made from petroleum sources outside the United States." At the same time, the ECA was admonished by Congress not to pay premium prices for foreign oil. Thus, Marshall Plan legislation had specific language which supported the policy goal of conserving western hemisphere resources by encouraging the development of Middle East oil.[51] As a Congressional study put the issue, "Only through the achievement of virtual balance of supply and demand within each of the Hemispheres can today's apparent petroleum shortage situation be overcome."[52]

The importance of the oil amendment cannot be understated. The United States had committed itself to European recovery but was unprepared to export domestic oil supplies in support of that goal. As a result, oil had to be procured in the Middle East. This, in turn, would require Washington to cultivate close diplomatic relations with oil-producing states, a task complicated by the Palestine conflict. Congress, however, reasoned that increased purchases of Arab oil would bolster local economies, leading producers to separate their economic and political preferences. According to political scientist Hadley Arkes, the petroleum legislation "tried to embrace all these considerations."[53]

The ECA was therefore given the go-ahead by Congress to assist the overseas operations of U.S. oil companies. These firms needed steel and specialized equipment to drill wells and build pipelines, refineries, and port facilities in the Middle East. In order to obtain such goods, export permits had to be issued by the Commerce Department, and ECA and State Department officials found themselves lobbying on behalf of the firms for export licenses.

One project of particular interest to both government and industry was the Trans-Arabian Pipeline (TAPLINE), which would ship oil from Saudi Arabia to the Mediterranean. Policymakers naturally linked completion of the pipeline to European reconstruction. "Under the Marshall Plan," the National Security Resources Board noted, "we will assume the responsibility for Western European recovery and economy and the completion of this line will make more oil available."[54] Indeed, as Defense Secretary Forrestal had feared, the Palestine conflict resulted in the closure in early 1948 of the oil pipelines running to Haifa, making new routes to the Mediterranean even more critical.

Forrestal was an active supporter of the TAPLINE. He stated that it was more important to ship steel to Saudi Arabia than to Texas. He told his diary:

> I took the position that because of the rapid depletion of American oil reserves and an equally rapid rising curve of consumption we would have to develop resources outside the country. The greatest field of untapped oil in the world is in the Middle East. . . . We should not be shipping a barrel of oil out of the United States to Europe.[55]

Secretary of State George Marshall also added his influential voice to the debate, writing Commerce Secretary Sawyer that the "oil of the Middle East is an important factor in the success of the European Recovery Program."[56] These "high policy" arguments in favor of steel exports convinced Commerce to grant the needed permits.

The involvement of senior government officials in decision making over steel export permits suggests the high policy importance of petroleum during the Marshall Plan period. U.S. policymakers recognized that ample oil supplies must be made available for European recovery; at the same time, supply patterns had to reflect U.S. and European security interests. By providing Europe with Middle East oil, and by conserving western hemisphere supplies, Washington could pursue its international economic and security objectives. Thus, even though Europe would become dependent on a politically volatile region for its fuel, American spare capacity would be available during an emergency. As the U.S. Congress pledged, U.S. oil could make up "the major share of any . . . world-wide deficits."[57] In sum, the United States had developed a new hemispheric oil policy of "West for West" and "East for East," with U.S. reserves acting as a strategic stockpile for alliance needs. This policy dovetailed neatly with the postwar marketing strategies of the American-based multinational oil companies.

But oil policy also had to react to changing market conditions. During the early Marshall Plan years, world oil supplies were tight, leading American officials to discourage CEEC consumption plans.

By early 1949 the world supply situation was dramatically different. ECA reported that U.S. demand, which represented 60 percent of the world total, had leveled off because of economic recession and a mild winter. At the same time, major additions were being made to global production and refining capacity. Incredibly, the world was now facing a glut, one that threatened to engulf America's "dollar oil" firms. This surplus precipitated a major Anglo–American conflict—the sterling–dollar oil problem—which posed yet another policy dilemma for U.S. officials.[58]

The Sterling–Dollar Oil Problem

With oil problems easing somewhat, countries could now choose between American and British suppliers. Given the dollar shortage, the vast majority of states clamored for oil that could be purchased with sterling. In that way, Marshall Plan funds were conserved for items that could only be bought in the United States. Further, Britain had accumulated debts around the world since 1939, and it was glad to pay these off by supplying petroleum.

British policies encouraged the world's appetite for sterling oil in three ways: first, by promoting bilateral trade agreements that gave sterling oil exclusive markets; second, through currency restrictions that limited the amount of dollars "sterling bloc" nations (i.e., much of the Commonwealth) could spend; and third, through trade and financial restrictions that limited American oil company activities in the British Empire. American firms were permitted to sell oil for sterling, but the British would not convert their sterling earnings back to dollars. Since dollars were used to pay for royalties, salaries, and equipment, the accumulation of sterling balances was of no interest to American firms. The sterling–dollar problem posed a severe challenge to the overseas operations of American multinationals.[59]

The British justified these measures as imperative to their dollar-saving strategy. Petroleum transactions were a major reason for the sterling bloc's dollar demand, and now that sterling oil was available, its use should be encouraged. Britain argued that it had no choice but to ration the few dollars left in its coffers.

The execution of British policy was bound to erode U.S. market share abroad, and oil company officials implored the ECA to respond. But the oil problem posed a dilemma for Paul Hoffman. If he sided with the American firms and got Britain to desist, he would not only subvert the recovery goal of easing the dollar shortage, he would also weaken a critical ally. If he backed Britain, he would contradict America's liberal economic objectives while harming the interests of important private sector actors. The ECA administrator faced a dilemma, and he found it difficult to formulate a policy that reflected enduring U.S. interests.

As the ECA pondered its options, the British went on the offensive. During the spring of 1949 exclusive sterling contracts were signed with Argentina, Egypt, and the Scandinavian countries. This action prompted Jersey Standard treasurer Leo Welch to write Paul Nitze of the State Department's ERP staff (Nitze happened to be, among other things, an heir to the Standard Oil fortune). Welch explained that "many traditional markets for dollar petroleum products and crude may, during the next three to five years, be drastically reduced or eliminated." If the market position of American firms continued to deteriorate, they would be sitting on Middle East oil that could not be sold. Arab sheiks, in turn, might consider giving these concessions to the British. Welch argued that it was a matter of national security to "support the American petroleum industry in discussions with the British."[60]

In September 1949, an Anglo–American ministerial meeting was held in Washington ostensibly on the topic of Britain's financial position; parallel talks were held on the oil problem at the official level. Discussions continued intermittently until December 20, but no agreement was reached on the oil controversy during this time. The Americans threatened to halt financing of Britain's refinery program, but the delegation did not budge. The British stated that there was no alternative to curtailment of American oil sales.[61]

In fact, the British emplaced additional discriminatory measures while the bilateral talks were under way. In mid-December the ECA's George Walden reported from Paris that

> the British have just informed affiliates in the U.K. and sterling area of American oil companies that import licenses will be denied them begin-

> ning January 1, 1950 for any products which they have customarily received from their parent suppliers, which can be in the future supplied from the surpluses of British oil companies.

American firms were now relegated to being distributors of sterling oil. What would they do with all that crude sitting beneath Saudi Arabia?[62]

Paul Hoffman finally acted. On December 19 he cabled the OSR to announce that "ECA will not consider at this time for financing British or British/Dutch company owned refineries in the U.K., on the continent or overseas. . . ." This was the first and only time that the ECA sanctioned a participating country during the Marshall Plan. Still, many congressmen felt that Hoffman had not gone far enough and urged him to cut all ERP aid to Britain.[63]

In Paris there was also strong sentiment for more decisive action. U.S. officials had long harbored suspicions of British oil policy, perceiving that London sought to control world oil supplies.[64] OSR economist Hollis Chenery suggested that Britain's aim was nothing less than "the invasion of existing and preemption of future American markets." George Walden added that the sterling–dollar problem was now "entirely out of the hands of the U.S. companies. Without fair and firm support from our government, the U.S. companies are helpless."[65]

With U.S. anger at the boiling point, London floated a proposal. The British said they would accept a reduction in, rather than exclusion of, dollar oil imports. If American companies pledged to purchase more goods and services in the sterling area, Britain would permit 9 million tons of oil imports per year over the next four years; the Americans had hoped to sell 13–15 million tons per year in the sterling bloc. In addition, the Americans would be permitted to convert only a small percentage of their sterling earnings into dollars.[66]

This proposal was unacceptable. The U.S.-based firms continued to seek arrangements that would permit them to compete on equal terms with their British counterparts. Sensitive to the sterling area's dollar shortage, various ideas were offered by the companies. Caltex (a Socal–Texaco joint venture that marketed oil in the eastern hemisphere), for example, argued that the British Treasury should provide the same sterling–dollar conversion services to both American and British companies. Anglo–Iranian and Royal Dutch/Shell were permitted to convert up to 25 percent of their sterling earnings to dollars in order to pay suppliers and royalties. American firms, Caltex argued, should be given the same allocation. In this way, American and British firms would become indistinguishable for financial purposes.[67]

W. L. Faust of Socony–Vacuum had another suggestion. He thought that the oil-consuming countries should pay more of the dollar costs of their oil puchases; why should the British Treasury continue to control all dollar disbursements? But this proposal was unattractive to London, for it retained a strategic interest in maintaining a viable sterling area. If the adjustment costs of the oil problem were placed on the sterling bloc's members, they would no longer have an incentive to remain participants.

As a result of the sterling–dollar oil problem, U.S. multinationals began to move more foreign oil to American shores; there were no other outlets. In early 1949, imports stood at 750,000 barrels per day; by January 1950 they had climbed to 885,000 barrels. Not surprisingly, Congress was disturbed by this trend and aimed to stop it.[68]

Wedged between Congress and the British Government, the U.S. firms offered a new set of proposals to London. Jersey Standard indicated its willingness to expand the throughput of its major British refinery by 110,000 barrels per day, and the company ordered several tankers from British shipyards. The company also agreed to reduce its convertibility demands. In return, Jersey asked that gasoline rationing be ended, in order to enlarge the oil market for all competitors.[69]

An agreement with Jersey was now in reach, but some details remained. First, Britain limited Jersey's sterling–dollar conversion to 15 percent of sterling earnings. Second, the firm agreed to sell its share of Iraq Petroleum Company output for nonconvertible sterling. For its part, on May 26, 1950, British Minister of Fuel and Power Noel-Baker announced the end of gasoline rationing.[70]

An event halfway across the world gave the talks with the other American firms a sudden urgency. On June 25, 1950, North Korean troops stormed across the 38th parallel, and a war was on. The demands created by American mobilization ended talk of a permanent oil glut. By September, all the firms had accepted the

"Jersey agreement" with Britain. The American multinationals accepted British discrimination as the price of reentry into the sterling market.

Ironically, the ECA issued its official petroleum policy paper one day after the invasion of South Korea. The paper said that ECA sought an end to discrimination against dollar oil, but it understood Britain's need to save dollars. The Administration decided that refinery construction in Europe would be considered on a "case-by-case" basis.[71]

The British gained significant market share during the sterling–dollar oil conflict. In 1938, Europe imported 11.3 million tons of dollar oil and 3.9 million tons of sterling oil; in 1951 7.4 million tons came from dollar sources, and 27.5 million from sterling companies. Further, despite ECA sanctions, Europe's refinery capacity increased by 150 percent between 1947 and 1951, and as a result 75 percent of its imports were now in the form of crude oil, as compared with 25 percent in 1938. Of Europe's dollar oil imports, the ECA financed 56 percent of the total, for $1.2 billion. This figure represented fourth place on the list of American-financed commodity purchases and 10 percent of the overall cost of the Marshall Plan. During the European Recovery Program, Europe's overall use of petroleum increased significantly, climbing from less than 10 percent of prewar energy consumption to nearly 15 percent in 1951.[72]

The sterling–dollar problem provides a case study of some of the strategic dilemmas faced by postwar American officials. With the Marshall Plan, the United States committed itself to the economic recovery of its Western allies. European reconstruction, however, could not occur in the absence of discrimination against American goods and services. But to what extent were allies who maintained closed economic blocs an asset to the United States? Further, how far could the allies push the United States before domestic actors, via Congress, reacted? The outcome of the oil problem was clearly favorable to Britain. This suggests that, in the final analysis, the maintenance of a critical ally was of primary importance to U.S. officials. The American government was prepared to sacrifice corporate profits and market share if the result was a strengthened United Kingdom.

Conclusion

During the Marshall Plan, an oil policy based on the hemispheric concept of "West for West" and "East for East" was largely realized. This meant that the United States could conserve domestic oil resources for an alliance emergency in the event Middle East supplies were disrupted. By encouraging this oil supply pattern, U.S. policy sought to serve alliance interests in the postwar world.

American officials recognized the risks inherent in this policy. A 1950 State Department memo expressed the problem as follows:

> The threat of Communist aggression is increasing. The Middle East is highly attractive and highly vulnerable to this threat. Rupture of the flow of Middle East oil to normal markets or seizure of those resources from without or within would seriously affect U.S. and allied economic, political, and strategic interests. . . .[73]

American power in the oil issue area was severely limited by the external challenges emanating from both friend and foe. Britain, the Soviet Union, and regional actors in the Middle East were all potentially disruptive of U.S. foreign policy. At the same time, the domestic oil industry in the United States placed severe constraints on policymakers; it would not permit, for example, high levels of imports. American officials had to formulate an oil strategy in light of numerous constraints.

Marshall Plan oil policies were developed in a context that accounted for U.S. and alliance economic and security concerns. The strategy that was adopted meant government support for rapid expansion of Middle East petroleum resources, coupled with conservation of domestic reserves by curtailing exports. This policy appeared to satisfy both domestic and alliance interests. In the following chapters, the durability of that strategy will be assessed.

4

Oil, Allies, and Iran: 1951–1954

Never had so few lost so much so stupidly and so fast.

DEAN ACHESON[1]

During the Marshall Plan years the United States became committed to an alliance oil policy based on hemispheric self-sufficiency. This geostrategy dictated that European recovery should be fueled with Middle East oil while American resources were conserved for domestic use and wartime emergencies. The hemispheric policy satisfied U.S. foreign policy interests in a manner acceptable to such powerful domestic actors as independent oil producers.

But Middle East oil flows were threatened by regional conflicts and the possibility of external attack. At the war's end, nationalist movements began to gather strength in oil-producing states, particularly in Iran and Iraq. The economic platforms of these movements may have varied, but all called for greater revenues from oil concessions. Another source of Middle East tension was the conflict between Arab and Jew, and after 1948 the United States had to walk a delicate line between its support for Israel on the one hand and its strategic dependence on Arab oil on the other. The great external threat to the region was posed by the Soviet Union which, in Harry Truman's words, was like a "sitting vulture on a fence waiting to pounce on the oil."[2]

This chapter deals with the first serious alliance oil crisis, caused by the nationalization of the Anglo–Iranian Oil Company (AIOC) by the Iranian Government in 1951. During this crisis, the United States sought to formulate a policy that would simultaneously enhance its prestige in the Middle East, bolster its leadership of the Western alliance, prevent possible Communist gains, and maintain oil flows. The United States was constrained in Iran, however, by its fundamental interest in maintaining a strong partnership with Great Britain. According to an agreement struck between Washington and London after World War II, Britain assumed responsibility for Middle East defense in the event of a global conflict with the Soviet Union. In effect, Iran was in Britain's sphere of influence. Although Washington found many aspects of British policy in Iran to be objectionable, it ultimately supported the approach to crisis management adopted by its staunchest Western ally.

The Iran–AIOC Dispute

The dispute involving the Iranian government and the Anglo–Iranian Oil Company had its roots in the various profit sharing agreements made between oil firms and the governments of oil-producing states during and after World War II. In 1943, Venezuela signed an agreement with its resident oil companies that gave it a "50/50" split in the profits arising from the sale of oil. This precedent was not lost on other states, and they negotiated contracts on similar terms. In 1949, in response to Iranian demands for a Venezuela-type agreement, the AIOC and the Iranian government renegotiated the 1933 contract that had governed their relationship.[3]

The 1949 agreement boosted the royalty payable to the Iranian government, and it provided additional financial benefits. The Iranian Majlis (Parliament) debated the terms throughout the year, but by 1949 it still showed no inclination to give its approval. AIOC, an oil monopoly whose majority shareholder was the British government, had become the preferred target of nationalist forces within Iran. The company was criticized not just on economic grounds but also for its hiring practices and its treatment of Iranian workers. As State Department official Richard Funkhouser observed in 1950, "AIOC and the British are genuinely hated in Iran. . . ."[4]

Among the nationalists' demands were increased employment of Iranians at all levels, including the managerial ranks, the right to audit AIOC's books and oil export figures, and renegotiation of the concession at 5- to 10-year intervals. The moderate Iranian Prime Minister, Razmara, urged AIOC to adopt these terms and to begin making payments at the new rate, despite the lack of Majlis ratification. The company rejected the nonfinancial demands, however, and refused to make payments until a formal agreement was signed.[5]

The United States was concerned by this dispute from the outset. In September 1950, after initial talks with British officials, State Department oil adviser Funkhouser wrote in a memorandum that the government should "urge the Foreign Office to accept Razmara's demands." Funkhouser was primarily worried about replacing Iranian oil in the event of a disruption of AIOC activity and said that U.S. officials must

> emphasize the oil supply picture. According to our oil companies, loss of Iranian oil *could not be replaced*. The Korean War has already created shortages even with US, Venezuela, and Middle East producing at record levels. Europe has only six weeks stocks and . . . three months would be required to divert Western Hemisphere supplies to AIOC normal markets (if supplies could be found). Ministry of Fuel and Power representatives were alarmed at this situation and the lack of US plans to supply Eastern Hemisphere markets in such an Iranian emergency . . . we told these representatives in effect that we felt that (a) first responsibility lay with the UK Government to take corrective action in Iran, (b) the situation in Iran demanded exceptionally liberal treatment.[6]

But the British were not prepared to compromise with Tehran. In formulating a policy response, the British had to consider their myriad economic and political interests throughout the Middle East; Egypt, for example, was now demanding British military withdrawal from the Suez Canal zone. As Foreign Secretary Herbert Morrison stated, British acceptance of a concession agreement dictated by Iran "would destroy our prestige in the Middle East with disastrous results in Egypt and elsewhere."[7]

For the United States, the major objectives with respect to Iran were to prevent a Communist takeover of the country either by external attack or internal subversion, and to maintain Iranian oil flows. The AIOC refinery at Abadan was the largest in the world, processing 25 million tons of petroleum products a year. In 1950 Iranian oil fields produced almost 32 million tons of crude, or 6 percent of total world output. The oil company supplied fully one quarter of Britain's domestic petroleum requirements, and it shipped Western Europe nearly 8 million tons of oil a year. The Abadan refinery also had direct military importance, in that it was the major supplier of aviation fuel and fuel oil in the eastern hemisphere.

These products were essential to Western military and naval activity in the Indian Ocean. As the National Security Council concluded, "Loss of Iranian crude and refined petroleum products . . . would directly and adversely affect the security interests of the United States. . . ."[8]

The United States believed its objectives would best be served by a rapid and peaceful Iranian–British agreement. But the United States could exercise little leverage over London. On the basis of bilateral agreements reached after the outbreak of the Korean war, the British accepted responsibility for defense of the Middle East in the event of a global conflict with the Soviet Union. As Ambassador Philip Jessup explained to Secretary of State Dean Acheson following military talks in London, "The U.S. Joint Chiefs of Staff . . . consider the area to be a British Commonwealth responsibility and will be unable to commit forces to that area during, at least, the first two years of war. . . ."[9] In short, Middle East security was a British problem.

In late 1950 the hard-line stance taken by AIOC suffered a significant blow as the American oil company ARAMCO signed a "50/50" profit-sharing agreement with Saudi Arabia's King Ibn Saud. According to a contemporary observer, this deal, "by exacerbating a feeling of injustice in other Middle East countries . . . contributed to the crisis which first broke in Iran, spread to Egypt, and presently threatened to shake the Western position throughout the Middle East area."[10] The nationalists in the Majlis grew intransigent, arguing that nationalization of the AIOC concession was the only way to satisfy the legitimate demands of the Iranian people. In December 1950 the Majlis officially withdrew from further consideration the supplemental agreement negotiatied between AIOC and the Iranian government in 1948.

The pace of events in Iran now accelerated. In February 1951, the oil company offered to launch a new round of negotiations, "but this offer came too late."[11] Prime Minister Razmara had lost control of the government, and on March 7 he was assassinated by a religious fanatic. In the ensuing weeks, the Majlis passed a number of nationalization acts. Rioting occurred at the Abadan refinery, and on April 12 two British employees of AIOC were killed. A veiled British threat of military operations followed, which further fueled the nationalist flames in Tehran.

In mid-April Assistant Secretary of State George McGhee met with British officials to discuss the evolving crisis. In his memoirs, McGhee recalled that the British "presented in forceful terms the significance they attached to the AIOC's concession. It provided 30 percent of Anglo–Dutch oil and £100 million annually to the British balance of payments. They did not wish to give up, weaken, or shorten the life of their concession." McGhee responded that Washington's first concern was "to prevent the loss of Iran to the free world. . . ." He reminded the British that oil nationalization was "a bitter pill" the United States had already swallowed in Mexico. He argued that Britain had no choice but to accept nationalization in principle and work out an agreement that would maintain an important role for AIOC in a new Iranian oil industry.[12]

State Department views regarding AIOC were also influenced by the sterling–dollar oil problem that had soured British–American relations during the Marshall Plan years. Iranian oil, it will be recalled, was used by Britain to displace "dollar oil" sold by American companies on world markets. American officials like McGhee who remembered these British oil tactics were hardly sympathetic toward AIOC's position.

On April 28, 1951, Dr. Muhammed Mossadegh, leader of the nationalist movement, became the new Prime Minister of Iran. By the end of the month an oil nationalization bill had passed the Majlis, and it was signed into law by the Shah on May 1. With this law, "the government was instructed to remove the AIOC from control of the oil industry."[13] Britain now recognized that a negotiated settlement was unlikely to be achieved.

Despite the nationalization bill, the United States would not condone precipitous British military action. On May 17, Secretary of State Acheson warned the British Ambassador to Washington, Oliver Franks, that the United States would support the introduction of British troops "only on invitation of the Iranian Government, or Soviet military intervention, or a Communist coup d'etat . . . or to evacuate British nationals in danger of attack . . ."[14] The United States did not wish to become entrapped in a military operation that would clearly be viewed as imperialistic throughout the Middle East.

The relationship between the United States and Iran at the time of nationalization was complex. As Richard Cottam has pointed out, "In 1951 pro-Americanism was one of the most striking features of Iranian nationalism. . . ."[15] The United States had won goodwill in Iran for its forceful action against the Soviet Union in 1946, and it was viewed as an enemy of British imperialism. Prime Minister Mossadegh entertained the hope that the United States would take advantage of its oil rivalry with Britain by stepping into the breach created by nationalization, striking a separate deal with Tehran which would permit American firms to operate the oil industry under contract.

On May 18, however, Secretary of State Acheson announced that no American oil company would displace AIOC in Iran.[16] To Mossadegh's disappointment, the United States would not exploit the Anglo–Iranian dispute on behalf of its corporations. If this left the United States without a positive policy, at least "American action . . . was directed at all times toward preventing irreparable steps by either Iranians or British, and thus leaving open the opportunity for common sense to reassert itself."[17]

Unwittingly or not, the United States had assumed the burden of mediator in the Anglo–Iranian conflict, though neither party had endowed Washington with any authority in the matter. In July, Dean Acheson dispatched foreign policy workhorse Averell Harriman to Tehran "to have a try at getting negotiations started again." Acheson feared that "jingoistic" pressures in Great Britain might force London to take military action, a move that he believed would invite Soviet retaliation. As Acheson saw it, "Armed intevention offered nothing except great trouble."[18]

Harriman arrived in Tehran on July 15 with the singular objective of getting Iran and Britain to the bargaining table. Although he told Mossadegh that the United States would be willing to consider Iranian requests for economic assistance, he rejected the notion of an explicit package tying U.S. aid to an AIOC agreement. Perhaps Harriman's greatest accomplishment in Tehran was to get the Shah to intercede in the oil dispute, something the ruler was hesitant to do given his "figurehead" status in Iranian politics. The Shah convinced Mossadegh that the time had come to negotiate directly with the British government. This was an important tactical decision for the Prime Minister, since previously he had stated that Iran's brief was with the oil *company*, and that the British government, while a major shareholder, had no direct standing in the affair. With Mossadegh's acquiescense, Harriman next traveled to London.[19]

Although the roving ambassador met with less than a cordial greeting in Britain, he nevertheless succeeded in getting the government to accept direct negotiations. Harriman suggested that the Anglo–Iranian talks be conducted in secret and that Britain appoint a delegate of cabinet rank. London's chosen representative was Lord Privy Seal Richard Stokes, who, according to Dean Acheson, proved not only clumsy in his negotiations with the Iranians but wont to make public the details of his discussions. Inevitably, the political parties in Iran would debate every issue that Stokes had mentioned, leaving Mossadegh no choice but to reject the British proposals. The Harriman–Stokes mission to Tehran ended in failure at the end of August.[20]

Alliance Energy Security

AIOC had scaled back its operations considerably since the rioting that had shaken the Abadan refinery complex in April 1951. By July, many AIOC personnel had left the country, and at the end of the month the refinery was shut down. With the collapse of the Harriman–Stokes mission, all British citizens working for AIOC quit the oil fields. The loss of Iranian crude and products posed the first major postwar challenge to alliance oil security, and, as Table 4–1 reveals, it occurred at a time of rapidly climbing European oil imports.

It will be recalled that as early as September 1950 the British had urged the United States to develop plans in the event of an oil emergency. The U.S. position was that Britain must take responsi-

TABLE 4-1. Western European Sources of Crude Petroleum, 1950–1952 (Millions of Metric Tons)

Source	*1950*	*1951*	*1952*
Indigenous production	3.9	4.9	5.9
Imports	39.5	56.6	71.9

Source: OECD, *Basic Statistics of Energy* (Paris: OECD, 1966).

bility to keep Iranian oil flowing. Nonetheless, the State Department promised to "speed up U.S. oil mobilization plans."[21]

These plans demanded close business–government cooperation. The U.S. Petroleum Administration for Defense (PAD), which had been established at the outset of the Korean war to allocate oil supplies among competing domestic users, invited the major oil companies to create a "Foreign Petroleum Supply Committee."[22] The members of this committee were authorized by the Secretary of the Interior to work together—and with their counterparts in Europe, including AIOC—to overcome the petroleum shortage. This meant they had to share corporate information and engage in cooperative practices in the areas of oil supply and distribution. As will be seen, this raised antitrust problems for the attorney general.

It proved relatively easy to replace Iranian crude with increased production elsewhere, notably Kuwait and Saudi Arabia. A more difficult problem was posed by the loss of jet fuel and other products from Abadan. Refinery capacity in the eastern hemisphere was limited, and U.S. policy was predicated on the conservation of western hemisphere capacity for domestic use and wartime emergencies. Further, tankers were largely engaged east of Suez and would have to be made available for the transatlantic run.

Energy crisis management also had a multilateral dimension. During the Abadan shutdown, the Oil Committee of the Organisation for European Economic Cooperation began to evolve institutional responsibilites for alliance energy shortages. As explained in Chapter 3, the oil committee's predecessor in the CEEC had been influential in formulating Marshall Plan oil policies and in shaping Western Europe's oil refinery program (the OEEC was created at the end of the Marshall Plan to continue the work of the CEEC).

Given the expertise of its delegates and the legitimacy it had gained as a forum for information sharing and energy planning, it was well suited to serve the allies in the event of an oil crisis.

The major concern of the oil committee's members was replacement of petroleum products from the Abadan refinery, which had supplied the OEEC countries with roughly 16 percent of their requirements.[23] While the committee did not play a role in allocating petroleum to member states, it provided information with respect to the plans being developed by the international oil companies in conjunction with the PAD in Washington and the Ministry of Fuel and Power in London. These emergency working groups reported regularly to the OEEC regarding product availability and allocation patterns. On the basis of this information, the committee could report that, while some temporary shortfalls might occur in specific markets, the overall impact of the Abadan shutdown would not be burdensome.

The oil companies had prepared well for the crisis. As a first line of defense, stocks had been built up and were at high levels at the time of the Abadan shutdown. Old refineries were returned to service, and new ones were expanded. By the summer of 1951, refineries in the western hemisphere and Europe were processing crude at full capacity. Aviation fuel was carted across the Atlantic, and only minimal shortages were experienced. The major problem that faced the companies was diverting tankers from their regular Middle East and Persian Gulf routes to the Atlantic. As the companies activated these emergency plans, consumer countries were urged by the OEEC to avoid any run on stocks or panic buying of oil.[24]

Since the fears of petroleum shortages passed quickly, the OEEC oil committee was not severely tested as a forum for energy crisis management. Yet the oil shock caused by the Abadan shutdown was not painless. Owing to the loss of Iranian crude and products, which had sold for sterling, Europe suffered a drain on its dollar reserves. In response to the Abadan riots of April 1951, oil prices climbed by $0.15 per barrel, and they increased another $0.15 before the year's end. By early 1952, prices were $0.40 above their 1950 level.[25] As Winston Churchill said of the financial strain:

> Now that the Abadan refinery has passed out of our hands we have to buy oil in dollars instead of sterling. This means that at least 300 million

> dollars have to be found every year by other forms of exports and services. The working people of this country must make and export at a rate of 1 million dollars more for every working day in a year. This is a dead loss which will affect our purchasing power abroad and the cost of living at home.[26]

But the strain was eased by the United States. Historian David Painter has pointed out that "in early 1952 the United States gave Britain $300 million in financial assistance to help cover the dollar costs of replacing Iranian oil."[27] Given Britain's critical role in Western defense, the United States was unwilling to take the risk that economic problems would cause London to trim its overseas commitments.

From an energy security perspective, the major lesson drawn by the OEEC during the Iranian oil crisis was the need to accelerate European refinery expansion and new construction. With the closure of Abadan, Europe had lost its major supplier of jet fuel and fuel oil, which could only be replaced by western hemisphere sources. Clearly, there was no shortage of crude oil in the world, but imports of products from "dollar" refineries were prohibitively expensive. Further, refineries were an obvious "bottleneck" in the oil system. Crude could be produced in many countries, but there were far fewer eastern hemisphere refineries. Petroleum products could be held hostage by those who controlled the refineries. To the OEEC, additional refinery capacity appeared to offer Europe its first line of defense in a future crisis.[28] National energy policies to encourage refinery expansion included the placement of high tariffs on imported petroleum products, and tax breaks for refinery construction.[29] Statistics reveal the OEEC's success in expanding the refinery sector. Whereas crude throughput in OEEC refineries totaled 57.5 million tons in 1951, it rose to 72 million tons in 1952 and 85 million in 1953.[30]

As suggested earlier, U.S. emergency planning was not without a distinctive twist. In 1952 the U.S. Congress released a Federal Trade Commission report, *The International Petroleum Cartel*, which provided evidence that prior to World War II the major oil firms had engaged in commercial practices contrary to U.S. antitrust law.[31] On June 23, President Truman ordered a grand jury investigation of the U.S.-based multinational oil companies, raising the specter of criminal indictments. The prosecution of an antitrust case, however, flew in the face of crisis management that required the close coordination of oil company operations.[32]

A bureaucratic battle ensued pitting the FTC and the Justice Department against State, Interior, and the National Security Council. The latter group argued that an antitrust exemption must be made, since "everything possible needed to be done to meet the shortages resulting from the closing of Iranian oil. . . ." Truman ruled that national security concerns dominated antitrust ones, and he permitted PAD to form a foreign supply committee "with broad powers to channel petroleum supplies to those areas short of petroleum." In 1953, Truman decided that both his judicial and national security objectives would be served by pursuing a civil as opposed to criminal case against the oil companies.[33]

Meetings with Mossadegh

The closure of Abadan, the necessity for Washington to work with London in meeting Europe's oil requirements, and Mossadegh's apparent intransigence caused the United States to reassess its Iran policy in 1952. The most recent bilateral U.S.–Iran talks, held in the United States in late 1951 following an address by Mossadegh before the United Nations in defense of the AIOC nationalization, had ended in failure. Further, the Conservatives, behind Winston Churchill, had been returned to power in Britain in October 1951, and Iran was a prominent campaign issue.

At the same time, the Iranian economy was in a state of collapse. With the AIOC closure, Iran had lost its primary source of foreign exchange. The British further exacerbated Tehran's economic problems by forbidding the Bank of England to exchange any Iranian earnings of pounds sterling into dollars, thereby depriving Iran of hard currency; it should be recalled that sterling was not freely convertible at this time. In addition, the major oil companies agreed to boycott Iranian oil, and the British impounded tramp tankers that left Abadan with petroleum products. As the economic problems mounted, the new U.S. ambassador, Loy Henderson, sought "to convince the shah to exercise his prerogative as monarch and replace Mossadegh with someone more "reasonable' on the oil issue."[34] Secretary of State Acheson meanwhile urged the British to prepare a "practicable plan of settlement" that could be presented to a post-Mossadegh regime.[35]

But the wily Mossadegh managed to emerge from this increased pressure with enhanced powers. The Tudeh (Communist) Party had thrown its support behind the prime minister and his nationalist party; previously the movements had been at odds. Perhaps it was British fear of Soviet intervention in the event of military action that led London at this point to suggest that Prime Minister Churchill and President Truman make a joint approach to Iran. The United States, still fishing for a peaceful solution, accepted the idea. The joint proposal floated to Tehran had as its main elements emergency economic aid for Iran, nationalization of AIOC with compensation to be determined by an international commission, and British purchase of Iranian oil at a discount. A series of clarifying letters between Tehran and the Anglo–Americans followed, but to no avail.

The attempt to reach agreement on the basis of the joint proposal was complicated in October 1952 by the appearance in Tehran of W. Alton Jones, president of Cities Service Oil Company in New York. Jones, to the embarrassment of the U.S. government, announced that he was prepared to recruit experts for the National Iranian Oil Company (NIOC) and to transport Iranian oil. He vowed that the British would not dare impound *his* tankers. While Mr. Jones only made a cameo appearance in the Iran drama, his actions undermined the perception of a united Western oil company front against Mossadegh. Emboldened by Jones, and upset by the British campaign against him, Mossadegh broke diplomatic relations with Great Britain at the end of October.[36]

The growing power of the Tudeh Party in Iranian politics, the diplomatic break with London, and the failure of Britain to offer reasonable solutions caused the United States to consider "one more big effort" aimed at a peaceful settlement.[37] With the collapse of the joint Anglo–American proposal, the United States had little choice but to pursue another tack. But, as Dean Acheson recounted in his memoirs, "An independent American solution over British opposition was more easily said than done." A unilateral policy could cause "damaging ill will . . . between our two countries," and in any event the United States had to cooperate with Britain on meeting oil shortfalls in Western Europe. During the Iranian dispute, the United States found its room for maneuver severely constrained, owing to its interest in maintaining a strong Great Britain.[38]

The United States played on this round its oil company card. Washington had hitherto discouraged American firms from purchasing Iranian oil or making deals at the expense of Anglo–Iranian. Acheson now reasoned that the hint of a policy change and the fear of American corporate entrance into Iran would move the Foreign Office to stay on the negotiation route. The Acheson proposal included a substantial loan to Iran against future oil deliveries and the purchasing and marketing of Iranian oil by a consortium of American companies, acting alone or in concert with AIOC.[39]

Acheson met with the oil companies to discuss the proposal in early December 1952. They expressed concern about an arrangement that secured better terms for Iran than other oil producers. They also told the Secretary of State that they "disliked even the appearance of hovering like vultures over the carcass of Anglo–Iranian." Clearly, the oil companies were not interested in pursuing a unilateral policy in Iran. Acheson later wrote that the meeting "reinforced my belief in the vast importance of joint action with the British. . . ."[40]

Acheson had guessed correctly that the intimation of American oil company participation in Iran with or without AIOC would catch Britain's attention. Accordingly, he and Foreign Secretary Anthony Eden hammered out the details of another package that U.S. Ambassador Henderson would bring to Mossadegh. Henderson met with the prime minister on Christmas Day 1952, and he left with cause for hope. Mossadegh accepted the idea of arbitration on the compensation issue, and of course welcomed an American loan payable against future oil deliveries. Unfortunately, a follow-up talk the next week left Henderson pessimistic. Mossadegh and the British still could not iron out their differences over the terms of compensation; would it be limited to AIOC's physical assets in Iran, or would it include AIOC's lost cash flows as a result of nationalization? The British were adamant on the latter point, and as a result Mossadegh rejected the Acheson–Eden proposal.[41] The belligerents had reached another dead end. For Acheson, time had run out. On January 20, 1953, Dwight David Eisenhower became President of the United States.

Eisenhower and Mossadegh

Mossadegh hoped that the new administration would differ from its predecessor with respect to Iran. After the November election he wrote the president-elect and told him of Iran's growing economic problems. He criticized the allegedly pro-British stance that the Truman administration had taken. Eisenhower replied that he wished for a bilateral relationship "characterized by confidence and trust."[42]

But the United States was already examining a radical alternative. In late 1952, British intelligence had invited CIA officer Kermit Roosevelt to London to propose a joint operation aimed at Mossadegh's ouster. Upon hearing Roosevelt's report, CIA Deputy Director Allen Dulles suggested that further planning await Eisenhower's inauguration; he was certain that Acheson would veto a covert operation. In February 1953, a few short weeks after the inauguration, representatives of British intelligence came to Washington to discuss "Operation Ajax" with now CIA Director Dulles and his brother, Secretary of State John Foster Dulles.[43]

On the basis of his long-standing military experience, President Eisenhower came to office with a nuanced view of the role of military power in foreign policy. Notable is the clear distinction he drew between overt and covert operations. Eisenhower believed that overt military power could only succeed when supported by public opinion and the Congress. He was also sensitive to world opinion and the dictates of international law. Covert operations, in contrast, provided a justifiable method for stopping Communist subversion in countries vital to U.S. national security. The very secrecy that surrounded covert action precluded the executive branch from seeking public support. Furthermore, covert operations generally required minimal resources. Eisenhower was anything but reckless in his use of force, and he carefully weighed the pros and cons of a CIA-sponsored coup against Mossadegh.[44]

As Iran's economic situation deteriorated in the spring of 1953, Mossadegh sent renewed pleas for help to President Eisenhower. He said that the international boycott of Iranian oil had caused Iran "great economic and political difficulties." Indeed, Mossadegh was digging into government pension funds to meet government expenses. But Eisenhower rebuffed the prime minister. "It would not be fair for American taxpayers," he said, "to extend any considerable amount of economic aid so long as Iran could have access to funds derived from the sale of its oil and oil products if a reasonable agreement were reached." The United States also criticized Mossadegh for his tolerance of the Tudeh Party.[45] Nonetheless, during the first half of 1953 the United States provided Iran with $23.4 million under its "Point Four" program that supported infrastructure and agricultural projects in the developing world.[46]

Meanwhile the political situation in Iran grew more disturbing. Mossadegh appeared to rely increasingly on the Tudeh Party for support, and fear of a Communist coup began to occupy Secretary of State Dulles. The British, sensing the growing dismay in Washington, now focused on the Communist threat. In meetings with U.S. officials, British intelligence downplayed the oil company dispute and placed the Iranian problem in the context of the East–West struggle. After three years, the British and American positions were beginning to fuse.[47]

According to Middle East scholar Barry Rubin, the U.S. government committed itself to a covert operation against Mossadegh on June 22, 1953. A small CIA team was assembled in Tehran over the following weeks, working in tandem with British operatives. On August 1, Roosevelt met with the Shah and briefed him on Operation Ajax. Upon learning that the mission was Anglo–American in character, the Shah reminded Roosevelt of Iran's strong "anti-British attitudes." Nonetheless, he pledged his support.[48]

Sensing a political meltdown in Tehran, Mossadegh launched a frantic effort to consolidate his powers. He appealed to the Soviet Union for assistance, perhaps hoping this would mobilize the United States to provide more aid. He then dissolved the rambunctious Majlis, in apparent contradiction of the constitution which vested this authority in the Shah. In response, the Shah's supporters took to the streets, only to be countered by the well-organized Communists. Tehran plunged into chaos as the mobs broke into shops, tore down statues, and overran government offices.

On August 12, with British and American backing, the Shah ordered that Mossadegh step down as prime minister, to be replaced by General Fazollah Zahedi. Mossadegh refused, and Tehran was rocked yet again by riots. The Shah fled the country, at which point the Anglo-American intelligence team organized a "massive demonstration" of his supporters.[49] Within a week the battle between the pro- and anti-Shah forces was over; on August 19 Mossadegh was arrested, Zahedi was restored as prime minister, and the Shah returned to Tehran.

As the Council on Foreign Relations remarked in its 1953 survey of world affairs, "to most Americans the fall of Mossadegh seemed . . . like a direct intervention of Divine Providence. . . ." But the council suggested the possibility that "agents of the Central Intelligence Agency also had a hand in the affair. . . ." Given that the council's vice-chairman (on leave) was CIA director Allen Dulles, such a comment might not have been idly placed in the annual survey.[50]

Yet while covert action against Mossadegh may have succeeded in its immediate objective, it also suggested the constraints on U.S. foreign policy. For two years, Dean Acheson had attempted to play the role of "honest broker" between Britain and Iran. The United States did not want to be identified with British imperialism, and it respected the sovereign right of developing countries to nationalize their resources so long as adequate compensation was paid. Britain, however, was able to maintain a hard line because of its valued role in the Western alliance. The United States would rely heavily on Britain in the event of a war with the Soviet Union, especially in the Middle East. Simply stated, British interests were more important to Washington than those of Iran.

Ironically, Britain also found that it had failed to achieve its policy goals. It had won U.S. support for covert action, and Operation Ajax was a complete success. A new man was in power, hand picked by Britain and the United States for his pro-Western views. But AIOC remained a hated symbol of imperialism, and it would not be permitted to regain its monopolistic position, by either Tehran or Washington.

The Iranian Consortium

With the fall of Mossadegh a new political structure, pro-West in orientation, was built in Iran. From an energy security perspective, the problem facing the United States and Britain was to reach a durable agreement with Tehran that would maintain the reliability of Iranian oil flows while avoiding renegotiation of every other Middle Eastern concession. The British Foreign Office initiated talks on this subject with the State Department immediately upon Mossadegh's overthrow.[51]

The British said that growing European reliance on Middle East oil posed a serious security problem. They suggested that the United States and Great Britain coordinate oil policy "to avoid whipsawing by the various producing countries." Such coordination would require that oil companies and governments work together, perhaps through an "international petroleum council" that would meet regularly. This proposal was reminiscent of earlier British efforts, both during and after World War II, to formulate an Anglo–American oil agreement.

The State Department viewed the Foreign Office proposal as seriously flawed. Domestically, it created antitrust problems. Internationally, the formation of a joint U.S.–U.K. oil commission might be viewed by oil producers—and other consumers—as a hostile act. Indeed, it might prompt the oil-producing states to form an organization of their own.[52] As State Department official Richard Funkhouser concluded, "It is thus more than possible that the results of United States and United Kingdom oil company collusion or cooperation . . . would produce something bad rather than good."[53]

Funkhouser, long familiar with foreign oil policy, recommended in contrast that "competitive forces" be encouraged to operate. He argued that one of AIOC's problems stemmed from its monopoly position, making it an obvious target for nationalist attack. He suggested that nationalization would have been less likely had several companies been active in Iran.

In October 1953, President Eisenhower appointed Herbert Hoover Jr. as a special representative of the United States with the objective of reaching a new Iranian oil agreement. Hoover's problem was to find a way of bringing Iranian oil back onto world markets in such a way as to maximize the Shah's revenues, a difficult task at a time of ample oil supplies. Indeed, the oil companies were concerned that the addition of Iranian crude and products on the market would lead to a price war, with declining revenues for every oil producer. According to a U.S. Senate study, "Hoover's solution . . . was to enlist the assistance of the five major American companies already operating in the Middle East to participate in an international consortium."[54]

In late 1953 and early 1954 Hoover negotiated a new oil agreement between Iran, the United States, and Great Britain. It called for a consortium of foreign oil companies to produce, refine, and market Iranian oil, with profits to be split between the firms and Tehran on a "50/50" basis. The National Iranian Oil Company (NIOC) would become owner of all properties once held by AIOC, with the concession to be renegotiated at twenty-five-year intervals. When viewed in toto, the consortium proposal managed to enhance Iranian prestige, break the AIOC monopoly, and reestablish Iran in world oil markets.[55] Former Prime Minister Mossadegh would live to see partly realized his vision of greater American influence in Iran at Britain's expense. The result was not a total loss for the British. AIOC won a 40 percent interest in the consortium, while the other 60 percent was divvied up among Royal Dutch Shell, Compagnie Française des Petroles, Standard Oil of New Jersey, Mobil, Socal, Gulf, and Texaco. The agreement was modified in 1955 when the U.S. administration sought to include a number of American "independent" oil companies in the consortium. This provided an important opening for smaller firms in Middle East concessions, breaking the stranglehold of the seven sisters.

With the signing of the consortium agreement, Abadan again bustled with activity. The new refining company overhauled most of the plant, and in 1955, 7.5 million tons of crude were processed. A substantial investment was made in housing, health care, and worker amenities. As part of the agreement, management of these ancillary activities was gradually taken over by NIOC. The oil industry was rapidly "Iranianized," and by 1956 there were only 453 British and American workers out of a total labor force of nearly 46,000. Perhaps most important to Tehran, oil revenues not only began flowing again, but they were markedly higher than preconsortium levels.[56] Overall, it appeared that Hoover had done his job well. Iranian oil would serve Western markets without interruption for twenty-five years.

Conclusion

During the Marshall Plan years, the United States encouraged its allies to increase their consumption of Middle East oil. This politically volatile region was in Britain's sphere of influence, and London had accepted responsibility for its defense in the event of a third world war. But Britain had various economic and political interests in the region besides containment of Communism. It believed that a hard line in Iran, backed implicitly by Washington, would best preserve those interests.

The United States attempted to manage the crisis by moving the belligerents to the negotiating table. This strategy satisfied the U.S. objective of containing Communism while promising the speedy return of Iranian oil to world markets. If successful, negotiation would also leave the British position in the region intact, if not enhanced.

Tehran under Mossadegh had yet a different set of objectives and accordingly made alternative calculations. At the beginning of his nationalization drive, Mossadegh thought that Anglo–American oil competition would lead the United States to cut a separate deal with the Iranians in which its oil companies would produce petroleum and operate the Abadan refinery. After this hope was dashed, he sought American aid, believing that Washington would provide economic assistance lest Iran fall into the Communist orbit. The aid came, but only in small amounts. With the passage of time, and little to show for his efforts, Mossadegh found it increasingly difficult to maintain a political coalition in Tehran. His turn to the Tudeh Party for support spelled his doom.

The energy security lessons of Iran for the Western alliance were more straightforward. It appeared relatively easy to make up the loss of crude oil from any single source, but the loss of petroleum products created serious logistical problems; clearly, Western Europe needed more refineries. The execution of emergency plans required government–business cooperation in Washington and London and multilateral support for the corporate allocation process, which was the responsiblity of the OEEC oil committee.

Overall, the Iranian dispute of 1951–1954 and associated oil shock demonstrated the resilience of alliance energy security measures. So long as the United States was willing to serve as the energy supplier of last resort, Western Europe need not fear dependence on Middle Eastern sources. The higher costs associated with "dollar" crude and products could be lessened by indigenous refinery expansion in Europe and, of course, by American financial assistance. At the same time, the maintenance of high stock levels could cushion the initial blow caused by a sudden oil shortage.

From the perspective of 1953, the overthrow of Mossadegh "justified . . . a measure of self-congratulation in Washington and throughout the Western world."[57] Oil was again flowing to world markets, and the Western alliance had maintained its cohesion. Of greatest importance, the alleged Communist tide in the Middle East had been turned, if not its nationalist undercurrent.

5

The Suez Crisis

In the Suez Canal the interdependence of nations achieves perhaps its highest point.

JOHN FOSTER DULLES, August 16, 1956[1]

Following his meetings in the United States in the autumn of 1951, Iranian Prime Minister Mossadegh returned to Tehran via Cairo. There he conferred with Egyptian leader Mustafa al-Hahas Pasha, who had recently ordered the British government to remove its troops from the Suez Canal zone, where they were stationed in accordance with a prewar Anglo–Egyptian treaty. In a speech to the Egyptian people, Mossadegh proclaimed that "all decisions concerning the Suez Canal or Iranian oil belong to the sovereign governments. Egypt and Iran share the same hopes and the same sufferings in all phases of their struggle."[2]

Just as Britain had feared, Mossadegh was taking his nationalistic message on the road. Ever since the outbreak of the AIOC dispute, the British government had told Washington that its policies in the Middle East must be influenced by the many political relationships and economic interests that it had throughout the region. Any policy that it developed with respect to Iran must take into account bilateral relations with Iraq, Jordan, Egypt, and the other countries in which British influence prevailed. Britain believed that hard-line policies were its best line of defense against indigenous nationalist movements.

But nationalist sentiment in the Middle East could not be easily contained. In Iran the hated symbol of Britain was the oil monopoly AIOC. In postwar Egypt it was the Suez Canal. Both these symbols were shrewdly manipulated by politicians who were determined to rid their countries of the European colonial heritage.

This chapter examines the Suez crisis of 1956–1957. As the story unfolds, many parallels with the Iranian case will emerge. In both cases, Britain faced powerful indigenous forces. In Egypt, as in Iran, British assets symbolized an imperial past. Both countries were in Britain's sphere of influence, and both were deemed of strategic importance in the Cold War.

But the differences were profound. Unlike Iran, Egypt was enmeshed in conflict not just with Britain but also with Israel and France. Unlike Iran, there was no strong internal Communist movement that threatened to take over the government. Further, unlike in Iran, there was no clear alternative to those in power. Of greatest importance, Britain and France would use military force at Suez to regain control of the canal, without the consent of the United States. Indeed, Washington feared that the militancy of its allies would open the Middle East floodgates to the Soviet Union. Owing to these differences, the nationalization of the Suez Canal Company in 1956 led to the most divisive crisis in postwar alliance relations.

From an energy security perspective, the oil supply disruptions caused by the closure of the Suez Canal in November 1956 were much more severe than those resulting from the Abadan shutdown. As will be seen, the European oil shortage that occurred during the winter of 1956–1957 could only be overcome with the active intervention of the United States government. But such assistance was purposely withheld until London and Paris met specific conditions set by Washington, including the withdrawal of their forces from the Suez Canal zone. America's willingness to employ economic coercion against Britain and France revealed the malign side of U.S. hegemonic power. This, in turn, prompted the allies to diversify their energy sources and suppliers, in an elusive search for security without dependence.

The Anglo-Egyptian Treaty

In 1936, as a response to Italian aggression in Africa, Britain and Egypt signed a treaty of alliance. Under the terms of the treaty, Britain was permitted to maintain bases and troops in Egypt, while the two countries jointly administered the Sudan. The treaty was for twenty years but could be revised prior to 1956 by mutual agreement.[3]

At the end of World War II, the Egyptian government indicated its desire to renegotiate the treaty, but the two sides proved unable to reach new terms. The British, who were responsible for Middle East defense under the alliance division of labor, could not agree to remove a substantial number of troops from the Suez Canal zone, which was viewed as a prime Soviet target in the event of hostilities. The United States also felt that a British military presence in the zone was important to the West's defense posture, and supported Britain in its talks with Egypt. George McGhee told Egypt's ambassador to Washington in July 1950 that it would not be wise to evacuate British troops from Egypt, since "Russian aggression in the Near East area was entirely possible and it would be essential to our common strategic plans to have the British on the spot."[4]

With the election of a supposedly moderate Wafd government in Egypt in 1950, Britain expected its concerns to be given a careful hearing. The Egyptians quickly disabused them of this idea, as British "occupation" of the Suez Canal had become a heated issue in the country. The Egyptians had taken the position that negotiations on the future of British base rights would be discussed only *after* London had given assurances that its troops would withdraw from the Canal zone and the Sudan. Cairo argued that British troops were no longer required, since the zone could be adequately defended in peacetime by Egyptian troops; the only item on the table concerned Britain's right to use the bases in the event of a global conflict.[5]

As the talks dragged on, British and American officials became worried that the simmering dispute might suddenly boil over into a larger conflict. In June the British mentioned the possibility of "scrapping the agreement" altogether and instead making a multilateral approach to Egypt that would include the United States.

The United States balked; the State Department regarded the problem as "one between the United Kingdom and Egypt." While Washington pledged "support" for the British position, it also believed that any new agreement must take into account the "national aspirations" of the Egyptian people.[6]

In October 1951 the long-feared crisis erupted, as the Egyptian government unilaterally abrogated the treaty of alliance. Britain immediately announced that it did not intend to respect this decision. Subsequently, fighting broke out between British troops and Egyptian "guerrillas," irregular forces sent into the canal zone with official encouragement. The British responded to one attack on their Suez base with a gruesome volley of heavy artillery that destroyed a police station, killing the lightly armed Egyptians, who were buried inside. The ensuing public outcry culminated in Black Saturday, January 26, 1952, when Egyptian mobs in Cairo sought vindication by wrecking European businesses and homes; the riots left seventeen Europeans dead.[7]

The ineffectual months of terror in Egypt only served to destabilize the Wafd government. It was neither able to dislodge the eighty thousand British troops stationed in the canal zone or to bring to justice Egyptian citizens who had crossed over from guerrilla to criminal behavior. Corruption in government was pervasive. Despite King Farouk's effort to contain the problem by subsituting one prime minister with another, the underlying malady remained. In July a coup d'état led by the Free Officers Movement under the titular leadership of General Nagib Bey overthrew the monarchy of King Farouk. The new military government brought order to the Egyptian capital, and negotiations with Britain were resumed.[8]

As the new governmental structure clarified, it became clear to Western observers that the mind behind the coup d'état was not Nagib but a young lieutenant colonel named Gamal Abdel Nasser. Nasser came from a middle-class background and as a teenager decided to devote himself to a military career after a brief flirtation with the law. He had fought in the 1948 war against Israel, and it was during this time that he joined a nationalist "cell" then forming in the army.[9]

The new military government appeared pragmatic in its policies and preoccupied with Egyptian economic development. This boded well for the West, a feeling heightened by the signing of an Anglo–Egyptian accord concerning the future of the Sudan in February 1953. The United States signaled its pleasure with the new turn of events by placing Cairo on the itinerary of Secretary of State John Foster Dulles's May 1953 swing through the Middle East. Dulles's visit became controversial after he presented a pearl-handled pistol to General Nagib, a gift from President Eisenhower. Throughout the Middle East, and in Britain as well, the question was asked: At whom is the gun pointed?

On July 27, 1954, Britain signed an agreement with the Egyptian government, now under the leadership of Prime Minister Nasser, which called for the withdrawal of British troops from the canal zone over the following twenty-four months. It also permitted British troops to return to Suez in the event of a war. Shortly after the agreement was signed, the United States provided Egypt with a grant of $40 million for economic development. According to one student of the Suez conflict, Herman Finer, Britain accepted the agreement "only after the strongest pressure . . . by Dulles and Eisenhower. . . ."[10] Finer asserts that the United States told Britain not to expect any military or political support from Washington in the event of another conflict with Egyptian troops in the canal zone.

As part of the Anglo–Egyptian agreement, Nasser pledged to uphold the 1888 Constantinople Convention that governed passage through the Suez Canal. Article I of the convention, which was signed by the nineteenth-century European great powers, stated the principle of free passage through the canal to all users. The powers had sworn "in the name of Almighty God" that "the Suez Maritime Canal shall always be free and open, in time of war as in time of peace, to every vessel of commerce or of war, without distinction of flag."[11] Since 1948, a notable exception had been made in the case of Israel, which was not given the right of free passage on account of the state of war that existed with Egypt. According to a State Department memo, Egypt was unprepared to extend this privilege unilaterally to Israel unless the other Arab belligerents were also ready to make peaceful gestures toward the Jewish state. While the United States made "several approaches" toward Egypt with regard to this issue, the Egyptians remained "extremely adamant in regard to lifting of restrictions. . . ."[12]

Although the Anglo–Egyptian treaty dispute was resolved peacefully, Britain recognized that the departure of its troops from the canal zone would have an enormous impact on its future position in the Middle East. For much of the twentieth century Britain had bolstered regimes in such countries as Iraq, Jordan, and Egypt with political, economic, and military support. After Israel's creation in 1948, London had adopted a deliberately pro-Arab stance. But neither history nor current policy was sufficient to preserve British military power in the region. And Britain was hardly aided in its efforts by Washington, which sought to distance itself from Europe's colonial heritage. Despite important responsibilites for Middle East defense, Britain had been left on its own to cut a deal with a nationalistic Egyptian government.

Arms and Aswan

With Britain retreating in the Middle East and U.S. policy toward the Arab world constrained by its support for Israel, the Soviet Union was given an unprecedented diplomatic opening. The new Soviet offensive was launched in September 1955, when Nasser announced that he would receive arms from Czechoslovakia as part of a barter deal that would send Egyptian cotton to the Eastern bloc. The Western powers, which had limited their arms deliveries to the Middle East under the "tripartite declaration" of 1950, recognized that Israel would have no choice but to seek new supplies in response. To U.S. officials, the Soviets were bent on destabilizing the Middle East in the hope of profiting from the ensuing chaos.[13]

The Soviet factor also emerged in Egypt's economic development plans. During the fall of 1955, while the repercussions of the arms deal were still being assessed, Nasser announced that he would build a high dam at Aswan, a project that the Egyptians had considered off and on for many years. In December he formed a High Dam Authority of prominent Egyptians who, in turn, invited consulting engineers from the West to study the project. At the same time, the Soviet Union pledged to provide financial support for dam construction.[14]

Secretary of State Dulles recognized immediately that Nasser was trying to play off East against West in order to secure the best financial package; such bargaining did not impress him. Nonetheless, the World Bank strongly supported the project, and by early 1956 Nasser had won a preliminary agreement from the United States, Britain, and the World Bank to finance the first phase of construction.

During the spring, Nasser's foreign policy took some turns that proved too drastic for Dulles to stomach. First, he announced that the Soviet Union had offered to build an atomic power station in Egypt. Then, in May, Egypt recognized the government of Communist China. In Dulles's view few acts could be more dastardly. He did not want to hear about the domestic political forces acting on Nasser that were apparently driving the Egyptian leader to signal his independence by dealing with both blocs. At this point Dulles began to question America's commitment to the dam.[15]

Nasser's relations with the West continued to deteriorate as Soviet Foreign Minister Shepilov visited Egypt in June 1956. The Russian diplomat joined Nasser in a celebration of Britain's final military withdrawal from the canal zone. In Washington, meanwhile, a congressional coalition representing cotton growers, Zionists, and anti-Communists had formed to fight the Aswan appropriation. On June 30, the appropriation for dam financing ran out and would have to await the next congressional session for renewal.

Recognizing that the dam had become a heated political issue in Congress, Egyptian Ambassador Hussein visited Secretary of State Dulles on July 19. He told Dulles that, if the United States did not finance the project, the Soviet Union would surely provide support. Despite being warned earlier by the French ambassador, Maurice Couve de Murville, that denial of the loan might mean nationalization of the Suez Canal Company, Dulles told Hussein that the United States had decided not to participate in the project at this time. This decision was not universally welcomed within the administration; Ambassador to the United Nations Henry Cabot Lodge called the denial "a terrible mistake."[16]

On July 26, 1956, President Nasser announced that Egypt was nationalizing the Suez Canal Company, and that its revenues would be used to pay for the Aswan High Dam. While some officials in alliance capitals had recognized this as a possible outcome of the loan dispute, the timing of the nationalization, and indeed the act itself, ultimately came as a surprise. It was assumed that the Egyptians were simply incapable of operating the canal without the support of the Europeans who managed its operations. With the nationalization, Britain's worst Middle East nightmare was finally realized.

The Suez Crisis

The Suez Canal was perhaps the nineteenth century's greatest feat of civil engineering. The canal, which joined the Red Sea and Mediterranean, opened on November 17, 1869, as a crowd of European royalty watched a great flotilla move south. The canal fulfilled the generation-long ambition of its animator, Ferdinand de Lesseps[17] (see Figure 5–1).

Although the canal occupied Egyptian sovereign territory, it was owned and operated by the Suez Canal Company, a private firm with headquarters in Paris whose shareholders were British and French; the British government was the largest single shareholder. Tolls were paid to the company, which in turn provided Egypt with a share of the profits. The initial concession granted by Egypt to the company was due to expire in 1968, at which time the agreement would be renegotiated. Nasser had pushed the clock forward twelve years and unilaterally decided on new terms.

For much of its history, the canal was viewed as Britain's gateway to India. But after India gained its independence in 1947, the canal was assessed more in terms of its economic contribution to European security. The quantity of goods moving through the canal in 1955 was 200 million tons, and of this cargo none was more vital than Middle East petroleum.

Middle East oil reached Europe via three routes: the canal and two pipelines with outlets on the eastern Mediterranean coast. Of this oil, 70 percent was shipped through Suez, the equivalent of 65 million tons per year. During the 1950s, Europe's reliance on Middle East oil was growing rapidly; 90 percent of supplies derived from concessions in this region. In 1955, the continent consumed 110 million tons of petroleum, and oil's share of the energy econ-

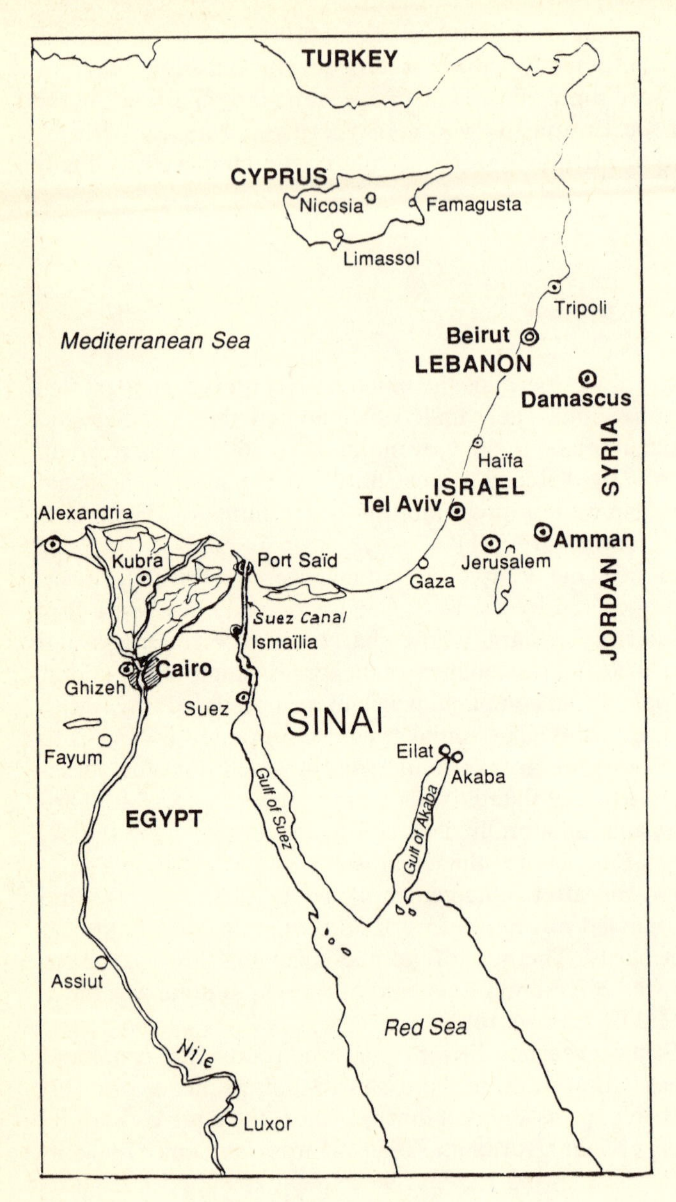

FIGURE 5–1. Sinai and the Suez Canal

TABLE 5–1 European Energy Consumption, 1950–1956
(In percentage)

Year	*Coal*	*Natural Gas*	*Petroleum*	*Hydropower*	*Total*
1950	82.5	0.3	14.0	3.2	100
1951	81.1	0.4	15.5	3.0	100
1952	79.6	0.5	16.7	3.2	100
1953	78.1	0.6	18.0	3.3	100
1954	76.8	0.8	18.9	3.5	100
1955	74.4	0.9	21.4	3.3	100
1956	72.7	1.0	22.9	3.4	100

Source: OECD, *Politique Energétique* (Paris: OECD, 1966), p. 31.

omy surpassed the 20 percent mark. Solid fuel was losing the price war with oil, and the coal industry was in a sharp decline (see Table 5–1).

The Europeans recognized that growing dependence on the Middle East was not without risk. In the wake of the Iranian shutdown, the OEEC oil committee expressed concern about the threat of supply disruption. But it was confident that western hemisphere supplies would be made available during any Middle East conflict.[18]

The seizure of the canal was greeted with panic in Britain. As the London *Star* editorialized, the canal was "an oil pipeline, an economic lifeline."[19] Prime Minister Anthony Eden said that Nasser had Western Europe by the "windpipe" and immediately ordered plans for a military assault.

But military action could not succeed without American support. Eden knew that Nasser would respond to a British invasion by blocking the canal, so American oil would be needed to make good any shortages. On July 27 the prime minister informed Eisenhower of his preparations, stressing that "the immediate threat is to the oil supplies of Western Europe. . . . If the Canal were closed we should have to ask you to help us. . . ."[20]

In response, Eisenhower immediately dispatched Under Secretary of State Robert Murphy to London; Dulles was in Peru. Murphy was told the British and French governments had agreed upon the need for military action to reclaim the canal. Eden asked the United States to play two roles in the event of a conflict: first, to provide a nuclear umbrella against any Soviet threat; second, to supply Europe with emergency oil supplies if necessary.[21]

Murphy made clear at the outset that U.S. political and economic interests diverged from those of Europe in the canal zone. The United States relied on the canal for only 300,000 barrels of oil a day, and it had no ownership stake in the canal company. From a political standpoint, the United States and Britain had differing views of Nasser. Although the United States had no reason to like Nasser on the basis of his recent behavior, Washington tended to view him as a pragmatic politician who would ultimately help to stabilize Egypt and the Middle East more broadly; he had from time to time indicated a willingness to negotiate with Israel. Although he had dealt with the Soviet Union and Red China, he was an anti-Communist and practicing Muslim who had put many Egyptian Communists in jail. To Britain, Nasser was an archrival who was determined to rid the Middle East of all British influence. As historian Geoffrey Barraclough has remarked, "Looked at from Washington . . . nothing was more likely to undermine American influence than for the United States to line up with the colonial powers against Egypt."[22]

The French brief against Nasser was not limited to the canal incident. Paris alleged that the Egyptian leader was providing military support to Algerian rebels who were fighting French rule. In France Nasser was portrayed as a Middle Eastern Hitler. To counter the Nasserian threat in the region, the French had begun in 1955 to provide Israel with advanced military equipment including jet fighters.[23]

For their part, the Egyptians had nationalized—not confiscated—the Suez Canal Company. Nasser had promised compensation to shareholders and had vowed to "safeguard the continuous unhindered flow of traffic through the canal." Since Nasser needed canal revenues to finance Aswan and his other economic development plans, it was unclear why Britain feared that he would close it and choke Europe's oil trade.[24]

President Eisenhower was deeply troubled by Murphy's report on the Anglo–French position. He urged Eden to use all available diplomatic channels before contemplating the use of force. He stated that military action would be viewed as illegal in the eyes of the international community, and that it would play into Soviet hands. According to a former CIA official who served in London at the time, Chester Cooper, Eisenhower dashed Eden's "hopes for American backing of early military operations. . . ."[25]

The U.S. position during the Suez crisis was not dictated solely by broad national security concerns. It must be remembered that the summer of 1956 found Dwight Eisenhower campaigning for reelection to the presidency on the twin pillars of "peace and prosperity." Chester Cooper has asserted that "domestic political considerations clearly dominated Washington's policy" during the crisis.[26] While "dominated" may be a strong word, Eisenhower feared that a Middle East war would not only damage his campaign but also those of Republicans involved in congressional races.

With the United States pledged to a diplomatic solution, Secretary of State Dulles called for a "maritime powers" meeting in London in August. But by way of contingency planning, Eisenhower also ordered the Office of Defense Mobilization to draw up emergency oil supply procedures in the event of a European shortfall. Washington was thus prepared to assist its allies in the event Nasser took precipitous action.

Emergency Planning

On July 31, five days after the Suez nationalization, the director of the Office of Defense Mobilization, Arthur Flemming, recommended that Secretary of the Interior Fred Seaton report that an oil emergency existed and that emergency plans should be established. Flemming noted that a foreign petroleum supply committee had been mobilized on an irregular basis since the Iran dispute, and that it was currently inactive. The committee, composed of major oil companies, was responsible for advising the U.S. government "on plans of action designed to prevent, eliminate, or alleviate shortages of petroleum supplies. . . which threaten to affect adversely the defense mobilization interests or programs of the United States."[27] The following day, Seaton approved the request and asked the attorney general to concur in order to avoid antitrust problems.[28]

By August 10, the Foreign Petroleum Supply Committee, with a membership consisting of such oil companies as Mobil, Texaco, and Standard Oil of New Jersey, had submitted a plan of action. The plan called for creation of a Middle East Emergency Committee (MEEC) composed of the major oil companies and several smaller "independents." In the event of an oil crisis MEEC would survey the supply/demand balance and make recommendations with respect to meeting any possible shortages. Activation of the emergency plan, which required the approval of President Eisenhower, would allow the companies to make any arrangements among themselves with respect to production, refining, transportation, and sale of crude petroleum and petroleum products. Administration oversight of the MEEC was ensured by the participation of a "full-time salaried Government employee."[29]

Operating parallel to MEEC in London was the Oil Emergency (London) Advisory Committee (OELAC), created by the British government "to advise on oil supply problems arising out of the Suez Canal emergency, and to collaborate with the MEEC."[30] OELAC was composed of officers from Royal Dutch/Shell, British Petroleum (successor to Anglo–Iranian), Compagnie Française des Petroles, and foreign-based executives of the American multinationals.

One important difference between emergency planning during the Iran dispute and the Suez crisis concerned the heightened responsibilities of the OEEC oil committee. It will be recalled that the oil committee served primarily as an information clearinghouse following the shutdown of Abadan. The closure of the canal, however, would create problems of far greater magnitude for Europe. Accordingly, after the nationalization, the OEEC council empowered the oil committee to begin preparation of oil allocation plans for Europe. The objective of this planning phase was to curb the panic that was growing within the member states. The OEEC warned that uncoordinated action by governments to secure petroleum supplies could "severely damage" the economic structure of Western Europe.[31]

Before any actual emergency existed, then, government sponsored structures had been created in the United States and Europe to cope with oil shortages. It was expected that, in the event the Suez Canal closed to tanker traffic, the shortfall would be made good by increased production in the western hemisphere. All that was required to set the mechanism in motion was President Eisenhower's command.

The Maritime Powers Conference

In preparation for the maritime powers conference in London that was to begin on August 16, President Eisenhower called a "bipartisan leadership meeting" that included congressional leaders of both parties as well as Vice President Nixon, Secretary of State Dulles, Chairman of the Joint Chiefs of Staff Admiral Radford, and Arthur Flemming of the Office of Defense Mobilization. Eisenhower told the congressmen that he sought a bipartisan policy on Suez since the outcome of the London conference might "be in treaty form."[32]

While briefing the officials, Dulles stated that the government had opposed the use of force against Nasser since this would give the Soviet Union "unbounded opportunities in the diplomatic sphere." But he said in the same breath that it was "almost intolerable" for Western Europe to rely on Nasser for access to Middle East oil. "This is almost a life and death issue for Britain and France," he claimed. The president chimed in that "Nasser's aggressive statements . . . seemed much like Hitler's in *Mein Kampf*. . . ." Dulles concurred, characterizing Nasser as "an extremely dangerous fanatic."[33]

Senator Lyndon Johnson said that the administration had only two courses of action: it must either "use all peaceful means to solve the problem" or else "tell our allies that we *are* their ally and support them." President Eisenhower said simply that "the U.S. will look to its own interests." The congressmen were too polite to ask exactly what those interests were, how they were ranked, and how they would be looked after.

The problem of European oil security was also discussed at the meeting. Dulles explained that two thirds of Europe's oil supplies traveled through the canal, while another one third was transported by pipeline. With anti-British feeling rising in the Middle East, he thought it possible that all these sources would be cut off in the event force was used at Suez. But Eisenhower doubted that Nasser would unilaterally close the canal, noting "the damage Arab nations would suffer if their income should be cut off." He said that "Nasser would begin to lose his prestige if he were responsible for cutting off oil income of other countries."[34]

Arthur Flemming then presented a detailed contingency analysis, on the basis of the plans developed by the Middle East Emergency Committee. He stated that, should the Suez Canal be closed, production in the Gulf Coast and the Caribbean could be increased "fairly soon" to meet Europe's requirements. If both the pipelines and canal were shut off, western hemisphere production must be accompanied by rationing in Europe. In either case, Flemming stressed that Europe would suffer an enormous "dollar drain" owing to the additional purchases of "dollar" oil; depending on the scenario the dollar drain was estimated at $400 million to $700 million. Senator Richard Russell reacted to Flemming's analysis by stating that "the real question" concerned U.S. financial support for Europe during a period of crisis. Eisenhower agreed and said that if Europe lost its access to Middle East petroleum, its "sheer existence" would be threatened.

The meeting confirmed in Dulles's mind the need for a negotiated settlement to the Suez dispute. Such a settlement would enhance U.S. prestige, avoid an alliance crisis, curb Soviet ambitions in the Middle East, and, possibly, win him the Nobel Peace Prize. With these thoughts, he set off for the maritime powers conference in London.

The twenty-two-power conference on the Suez Canal met from August 16 to 23, 1956; Egypt was notable by its refusal to attend. Dulles's objective was to present the Egyptian Government with a multilateral *fait accompli*, an agreement that the canal must be operated under international auspices.[35] This goal was eloquently stated in his opening speech:

> What is required is a permanent operation of the Canal under an international system. . . . Confidence is what we seek; and for this it is indispensable that there should be an administration of the Canal which is nonpolitical in its operation. . . . The Canal should not be allowed to become an instrument of policy of any nation or any group of nations, whether of Europe, or Asia or Africa.[36]

Dulles proposed the establishment of an international board, to be associated with the United Nations, which would operate the canal; Egypt, of course, was to be one of its members. Egypt would receive an "equitable return" on the profits generated by canal traffic. Dulles recognized Egypt's right to nationalize the Suez Canal Company but stressed that "fair compensation" must be provided all shareholders. In sum, the Dulles plan recognized both Egyptian sovereignty and the international character of the waterway.

The main dissenters to this approach were Ceylon, India, Indonesia, and the Soviet Union. They argued that two issues were at stake: first, the legality of Nasser's action of July 26; second, free passage through the canal. With regard to the first point, the delegates asserted that there was no question concerning Nasser's right to nationalize the Suez Canal Company. Regarding free passage, they took Nasser at his word that no interference with shipping would occur; it should be noted that neither Dulles nor his opponents mentioned Israel when they discoursed about the right to free passage. The Soviet delegate recognized that the maritime nations might be apprehensive about a sudden rise in tolls and suggested that "it would be desirable to discuss with Egypt the question of tariffs." But the four dissenters did not believe that the creation of an international board was justified by the circumstances.

The maritime conference ended on August 23 with eighteen nations, including all the Western allies, in support of negotiations with Nasser on the basis of the Dulles proposal. The Prime Minister of Australia, Robert Menzies, was selected to represent the group in Cairo, where he met with Nasser in early September. There were no great expectations for success. The Egyptian leader's hand had been strengthened by Soviet and Indian support for his position and by Egypt's ability to maintain canal shipping at normal rates. Time was on Nasser's side.

Nasser rejected the London proposals in a letter to Menzies of September 9, following their meetings. He vigorously defended his action and affirmed his country's devotion to "freedom of passage" through the canal. He also stressed that plans were being made to improve the canal so that larger tankers could be afforded transit. Nasser observed that freedom of passage was already guaranteed by Egypt's military presence in the canal zone (British troops of course had withdrawn), and he could not see how an international board could provide any additional safeguards. After reading the response, Robert Menzies simply reported to the eighteen powers that his task "was completed."[37]

With the failure of the Menzies mission, Britain again became committed to the use of force against Nasser. During the Cairo talks Eden wrote Eisenhower of his fear that

> the seizure of the Suez Canal is . . . the opening gambit in a planned campaign designed by Nasser to expel all western influence and interests from Arab countries. . . . [I]t will be a matter of months before revolution breaks out in the oil-bearing countries and the west is wholly deprived of Middle East oil.[38]

The Prime Minister had thus formulated a Middle East version of the domino theory.

But Eisenhower replied that negotiations must be continued and a peaceful settlement found. "Of course," he wrote, "if during this process Nasser himself resorts to violence . . . then that would create a new situation. . . ."[39] The president had tipped his hand; he would only support force against Nasser if the Egyptian leader unilaterally blocked the canal.

Early September found Dulles in yet another feverish effort to find a diplomatic solution. Now he proposed the creation of a Suez Canal Users' Association (SCUA) which would include at a minimum the United States, Britain, and France.[40] SCUA would employ pilots to direct canal traffic and assume responsibility for operation of the canal on behalf of all users. The justification for the proposal derived from the fact that several French and British pilots had left the canal company following the nationalization and that most of the pilots were now relatively inexperienced Egyptians. The Egyptian government was requested to "cooperate" with SCUA in order to maximize the flow of ships through the canal.[41]

The SCUA idea was greeted with widespread skepticism. As Geoffrey Barraclough wrote in his 1956 survey of international affairs, "Neither the legal nor the practical arguments in favour of the scheme were . . . very compelling."[42] Nonetheless, Dulles was able to win alliance support for SCUA by promising U.S. oil supplies in case the canal was closed by Nasser; the United States would also provide a *loan* to enable Europe to purchase any oil that was needed.[43] Yet another London conference was held in September, this one attended only by the eighteen supporters of the earlier Dulles proposal. But it proved difficult to muster enthusiasm for SCUA. As time went on and canal traffic continued unabated, "sympathy in Europe as well as in Asia was visibly veering to the Egyptian side. . . ."[44] The conference concluded on September 21 with several major parties, including France, postponing any decision about whether to join the association.

By now it appeared that Dulles was doing little more than buying time. But time for what? To CIA offical Chester Cooper he was obviously buying time "to get the Administration through the remaining sixty days of presidential campaigning without a Middle East War."[45]

Britain and France were also buying time: time to mobilize their military plan of action. During the first part of October, Paris and London agreed to yet another effort at a negotiated settlement, this time through the good offices of United Nations Secretary Dag Hammarskjold. Simultaneously, various resolutions concerning the Suez dispute were debated in the Security Council. After a Security Council vote on October 13 against a British–French resolution condemning the nationalization, the two allies agreed that only a military solution remained.[46]

An added dimension to the Suez crisis was introduced by growing tension between Israel and its Arab neighbors. Throughout the year terrorists had been entering Israel from Jordan and Gaza in order to carry out attacks on Israeli citizens. Israeli reprisals became increasingly violent. Facing a common enemy in Nasser, France and Israel developed a close bond that resulted in substantial shipments of advanced French military equipment to the Jewish state. By mid-October Israel had become a partner in the military operations now being planned by France and Britain against Egypt.

The Use of Force

From the middle of October onward, a "communications blackout" interrupted the normal flow of information between London,

Paris, and Tel Aviv on the one hand and Washington on the other. Suddenly, American diplomats were cut off from decision makers in those capitals. While U.S. intelligence observed an increase in communications between Paris and Tel Aviv and had become aware that Israeli forces were mobilizing, the scope and timing of operations against Egypt were uncertain.

The extent of coordination among Israel, Britain, and France remains a matter of academic dispute. While it is clear that Israel and France cooperated closely in scheming against Nasser, it is possible that Paris was unaware of the specific date when Israel would launch operations against Egypt. For its part, London maintained its distance from the Jewish state throughout the Suez crisis in order to minimize the damage to its bilateral relations with such allies as Jordan and Iraq. Indeed, if London influenced Israeli planning, it was to steer reprisals for terrorist attacks away from Jordan and into Gaza.[47]

A devious plan of action had been hammered out in the three capitals. First, in early November—possibly on election day in the United States—Israel would launch an attack aimed at terrorist targets in the Sinai peninsula and at Egyptian military bases around Sharm el Sheikh. As the Israelis advanced across the Sinai toward Suez, France and Britain would hand down an ultimatum: Egypt and Israel must cease fire, or else a joint operation would be launched to separate the belligerents and maintain shipping through the canal. After Nasser rejected the ultimatum, as he surely would, French and British troops would invade and capture the canal zone. It was hoped that Nasser would be disgraced and fall from office and that a new, more moderate government would be formed in Cairo. This was the common, if fantastic, objective of David Ben-Gurion, Anthony Eden, and Guy Mollet.

With the Soviet invasion of Hungary on October 24, Israel advanced its battle plans, to the surprise of its partners. Ben-Gurion calculated that the Kremlin now had its hands full and would be unable to intervene militarily to assist Egypt. At the same time, the Soviet Union would hardly be in a position to denounce Israel in the United Nations.[48]

On October 29, Israel launched its attack, supported by French airdrops of supplies. The Israelis moved quickly—faster than expected—and within twenty-four hours they knew their military objectives would be realized. When Dulles learned of the Israeli attack, he immediately called for a U.N. Security Council session. The next day, while the Security Council debated a U.S. resolution calling for a cease-fire, Anthony Eden unilaterally sent the belligerents an ultimatum. The French and British delegates then vetoed the U.S. resolution, arguing that their governments had already acted to stem the crisis. Ironically, the Soviet Union had supported Dulles's call for a cease-fire.

As part of the ultimatum, Egypt was asked to accept a "temporary occupation" of Port Said, Ismalia, and Suez by British and French troops; Nasser and Ben-Gurion were given twelve hours to respond. A State Department official called it the "most brutal ultimatum in modern history," probably suggesting his shock more than his lack of historical perspective. The ultimatum expired on October 31, and the same day British and French aircraft launched attacks against Egyptian airfields.[49]

The United States, along with almost every U.N. member, expressed outrage at the joint action. But further Security Council action was effectively stymied by British and French veto power. The debate thus moved to the General Assembly, and on November 1 an overwhelming majority approved a cease-fire resolution.

In the Middle East the situation deteriorated. On November 2 and 3, Egypt sank blockships in the canal, halting the flow of traffic. On November 4 sections of the Iraq Petroleum Company's pipeline that passed through Syria were blown up. Europe was now facing a severe oil shortage, but "it was made clear that America . . . was in no mood to help out."[50]

On November 5 Britain and France launched a paratroop and seaborne invasion of the canal zone. The timing was odd; Egypt and Israel had accepted the General Assembly's cease-fire resolution before the joint force landed. Eden and Mollet defended their "police action," stating that it was necessary to "stop the hostilities" and "separate the combatants." But the reasoning was specious in light of the situation on the ground. In the eyes of the world, the two European powers were simply invading Egypt.

The Soviet Union now threatened to take unilateral action. In letters to Eden and Mollet, Premier Bulganin asked what their reaction would be if they "were attacked by other states having at their disposal modern and terrible means of destruction." He continued in his letter to Eden, "If this war is not stopped it carries the danger of turning into a third world war." The Soviets followed up on this tacit nuclear threat by proposing to Washington that a joint Soviet–American force intervene in Egypt if hostilities did not cease.[51]

The Soviet Union played its cards well during the Suez crisis, serving both as Arab nationalism's protector and as the promoter of world peace; the latter role was a difficult one in light of the Hungarian episode. The Soviet offer of joint action was neatly calculated to entrap the United States; Washington would of course reject the idea and in so doing appear to support its allies' aggression. Eisenhower had to react in a way that would maintain U.S. prestige.

London and Paris were shaken by the Bulganin letter. Officials saw the Soviet action as an example of "nuclear blackmail" and trembled at the thought the Suez invasion might trigger World War III.[52] Worse, it was unclear how their unreliable American ally would respond.

Herman Finer has asserted that the United States also feared Soviet military action, whether in the form of an attack against British and French troops in Egypt, or a direct strike on London and Paris.[53] While the fear quotient is debatable, Eisenhower acted decisively to oppose the Soviets. He made clear that the United States would respond to any attack on U.S. allies; in the words of NATO Supreme Commander Gruenther, the Soviets would be destroyed "as surely as night follows day."[54] Military leaves were canceled, and U.S. forces were placed on various levels of alert. Despite this firm response, it should be noted that NATO's force structure had been weakened by the redeployment of French and British troops from the European to the Egyptian theater, creating yet another point of tension in alliance relations.[55]

As it countered the Soviet threat, the United States also placed severe economic sanctions on its allies. President Eisenhower refused to activate the oil emergency plan, telling an associate that "those who began this operation should be left to work out their own oil problems—to boil in their own oil, so to speak."[56] He also refused to support the French franc and British pound, which were now in virtual free-fall. The British and French saw that the costs associated with Suez were quickly rising, fueling political discord at home. Indeed, "Suez divided Britain more deeply than any other event since Munich."[57]

On November 6, Anthony Eden called a halt to the invasion of Egypt; Guy Mollet soon followed suit. According to historian Barraclough, Soviet threats had played their part in the decision, but "far more serious appears to have been the pressure from Washington. . . ." The costs of the Suez operation were enormous and growing. The military action itself had cost £100 million, while the additional oil bills were estimated at £60 million per month. Of even greater concern, British dollar and gold reserves were draining quickly as the pound fell; in November alone 15 percent of the total disappeared.[58]

The end of invasion did not bring an end to hostilities. Although a cease-fire was now in place, Israeli, British, and French troops remained on Egyptian territory. In the United Nations the details of an emergency force to assume peacekeeping duties were being hammered out. Meanwhile, Nasser took advantage of the lull in fighting to move arms into the canal zone and resume attacks on the foreign troops. As the fire fights became hotter in intensity, the allies began to reconsider the cease-fire, and they drew up a proposal to renew the attack. President Eisenhower reiterated that he would not activate the Middle East Emergency Committee until he received an acceptable troop withdrawal plan from London and Paris. By the end of November, with winter on the horizon and energy shortages imminent, the allies agreed to leave Egypt.

Eisenhower had won his gambit; he had successfully used economic pressure to coerce Britain and France. But it is interesting to speculate how far Eisenhower would have gone had Mollet and Eden called his bluff and maintained their Suez strategy. At what point would the threat of severe alliance economic problems have overcome U.S. opposition to the invasion? Would Eisenhower have been willing to sit and watch the European economy tumble? Dulles raised these questions tacitly with British Foreign Minister Selwyn Lloyd in a mid-November meeting, while the crisis still raged. "Well," he asked, "once you started, why didn't you go through with it and get Nasser down?" "Foster," Lloyd answered, "why didn't you give us a wink?" "Oh!" Dulles cried, "I couldn't do anything like that!"[59] One of the peculiarities of the Suez crisis was the ambiguity that characterized alliance relations from beginning to end.

Sharing Oil

With the loss of the canal and the IPC pipeline, Europe was faced with oil shortages of nearly 2 million barrels per day, or two thirds of its normal supplies. In addition, Saudi Arabia had embargoed the sale of oil to Britain and France. The continent thus faced an energy crisis unprecedented since the late 1940s.[60]

A solution to the crisis required increased oil production in the western hemisphere and the rerouting of tankers. According to the OEEC, the latter problem was one "of the very greatest complexity."[61] First, tankers that carted Persian Gulf supplies to Western Europe now had to be rerouted around the Cape of Good Hope, adding weeks to the voyage and, in the absence of greatly increased numbers of tankers, less oil for the continent. Second, if more oil was to be shipped from western hemisphere sources like Venezuela and Texas, it meant that tankers would have to be made available for transatlantic routes; these tankers must come either from the Middle East or from their regular U.S. coastal runs. Even if tankers could be found to handle these chores, it was clear that their owners would take advantage of the situation by increasing rates and making the oil lift an expensive proposition.

Faced with certain shortages, the OEEC oil committee, in cooperation with the emergency committees in London and Washington, created an OEEC Petroleum Emergency Group (OPEG), which would be responsible for oil allocations to OEEC member countries; OPEG's first meeting took place on December 6, 1956. Unlike crisis management during the Abadan shutdown, when the oil companies themselves made allocation decisions, an international organization assumed the responsibility this time. Given the severity of the oil shortages, the companies were happy to leave this task to an official body. In that way they might avoid becoming the scapegoats of European governments.

At the allocation meetings a detailed survey was made of each member country's supply situation. The oil companies were told to allocate supplies on the basis of a formula that took into account pre-Suez oil consumption, stock levels, and indigenous energy production. The committee strove to make oil allocations as equitable as possible so that the burden of shortfall would be distributed; its watchword was "equalisation." At the same time, a special 200,000-

ton oil reserve was set aside every ten days for emergency distribution to countries in particular need. Overall, OPEG "assumed the task of deciding how the available supplies for Europe were to be shared, with industry at the international and national level acting as the executive body carrying out the decisions of the Organisation and advising on the supply position."[62]

At the outset of the crisis Europe looked to Venezuela rather than Texas for its emergency supplies. The Lone Star state's contribution was limited since the amount of "allowable" oil production was determined by a rationing board, the Texas Railroad Commission (TRC). The TRC, under the strong-willed leadership of General Ernest O. Thompson, established the level of production for each of the state's 160,000 wells. Acting in the interests of the small producers, Thompson refused to boost allowable production despite MEEC requests; a tight oil market would result in higher prices. In January oil prices on the U.S. gulf coast climbed from $2.82 per barrel to $3.17.[63]

Thompson's policy changed only following a threat from President Eisenhower that he had the authority and was prepared to take over the TRC's functions. Texas production subsequently climbed from 3.3 million barrels per day in November 1956 to 3.5 million in February and 3.7 million in March. Eisenhower also acted to ease the energy shortage by donating some mothballed Navy tankers to the oil lift.[64]

That Texas oil production climbed quickly once the allowable quota was changed demonstrates that the January price hike was the result of TRC policy and not tight market conditions.[65] The ability of the commission to manipulate market conditions in favor of independent producers again exemplifies the American "paradox of external strength and domestic weakness." It also demonstrates how crises can present windfall economic opportunities for specific actors; recall how the United Mine Workers took advantage of Europe's postwar coal shortage. A presidential threat may have ended the TRC fray, but time and money were lost in the process.[66]

The Suez crisis led to a dramatic shift in Europe's oil sources. As Table 5–2 reveals, the United States became an especially important emergency supplier. The ability of shipowners to place more tankers on the Persian Gulf run is also attested to by the statistics. Overall, by the spring Europe was able to maintain pre-Suez consumption levels, if not exceed them.

TABLE 5–2. Movements of Petroleum to Western Europe (Thousands of Barrels per Day)

	Origin			
	U.S.	*Caribbean*	*Mideast Pipelines*	*Persian Gulf*
Pre-Suez	40	530	330	990 (via Suez)
Post-Suez	330	590	130	970 (via Cape)

Source: U.S. Congress. Senate, *Emergency Oil Lift*, pp. 106–107.

By the end of May, the oil crisis had ended with the reopening of the canal and IPC pipeline. The various emergency oil committees—MEEC, OELAC, OPEG—disbanded. At OPEG's final meeting on May 2, 1957, the group's chairman, an executive of Royal Dutch/Shell, complimented all participants on a job well done. Panic was avoided, he said, by "coordinated action between European governments and the international supplying industry. . . ."[67] In its official report on the crisis, the OEEC argued that "without the will to cooperate, at worst a mutually destructive scramble would have developed, while at best availability would have been unevenly spread. . . . The economic consequences of such confusion would inevitably have been severe."[68]

But, as with Iran, the crisis was not without its economic repercussions. Owing to a combination of oil price hikes, increases in tanker rates, and the shift from sterling to dollar oil, Europe faced an additional oil bill of $300 million to $400 million. These dollar costs meant a deterioration in Europe's balance of payments with the United States, which could only be corrected by increased exports. The overall impact of the shock, however, was buffered by the fact that the European economy had been enjoying a strong expansion at the time of Suez and by the ample energy supplies that were ultimately provided as the emergency plans were activated.[69]

After the crisis passed, the OEEC reflected on the lessons learned. One internal memo noted that

> the impact of the Suez crisis on the energy supply of Western Europe has confirmed the fears . . . that there are inevitable risks in the in-

> creasing dependence of Western Europe on outside energy supplies, particularly when most of them must come from one small area of the world.[70]

The oil committee recommended the following policy measures: (1) the accumulation of larger stockpiles in Europe; (2) efforts to make Europe's oil supply system more "flexible" by developing new tanker routes and modernizing refineries; (3) seeking new sources of supply; (4) mutual consultation and appropriate planning by governments and industry to anticipate future oil interruptions. Policy implementation was all the more important, since "Europe should not count on the indefinite continuation of the present export potential of the United States."[71]

Yet another suggestion floated at the time of the Suez crisis was creation of a "European Oil Community, designed to improve Europe's bargaining power vis-à-vis Middle East oil producing countries on a commercial as well as political level."[72] This organization would, in effect, have supranational powers in the oil realm. While such an idea was attractive in light of multilateral cooperation within the coal and steel community, an obvious difference was that Europe's oil supplies were located outside the continent, in a region where the European states themselves had diverse economic and political interests. Even if the will existed to form such an organization, it still looked like a poor solution to the energy security problem. As prominent oil consultant Walter Levy pointed out,

> there are . . . distinct trends toward a unified oil policy by the Middle East producing countries. . . . Nothing would give it greater impetus than a combination of European consuming countries. It is to Europe's interest, however, to avoid as long as possible the establishment of such a Middle East bloc. . . .[73]

For his part, Levy recommended an oil policy based on three principles: first, the "community of interest" between Middle East states and the West; second, minimization of Western vulnerability to Middle East oil interruptions using appropriate energy policies, including the creation of "strategic" oil reserves; and finally, governmental support of "the rights of private companies," which was fundamental to increased investment.

Yet despite the conciliatory tone in his first point, Levy was critical of U.S. policy during the Suez crisis. He wrote that the United States "cannot ignore certain vital interests of Britain and France which run counter to Arab national ambitions. . . . The maintenance of the NATO alliance imposes upon us the obligation to protect their vital interest as well as ours."[74] Levy had struck at a fundamental problem inherent in the Western alliance. The problem, which informs this entire study of alliance energy relations, concerns whether military allies can manage their conflicting interests without reducing their collective security.

Conclusion

The Suez debacle would continue to bring pain to alliance governments for months after the event. It brought about the downfall of Anthony Eden, and it exacerbated the Algerian problems of the French. While Ben-Gurion had realized many of his immediate military aims in the Suez campaign, ultimately he was forced to withdraw from the Sinai peninsula under pressure from Washington. In the United States, the Suez debacle catalyzed the formulation of new regional and alliance policies, dubbed "pactomania," including the Eisenhower Doctrine that gave containment a specifically Middle East flavor. Nasser and Khrushchev were perhaps the sole unblemished survivors of the crisis.

No aspect of the Suez dispute pained Eisenhower more than its alliance dimension. He was a firm believer in the Western alliance and, according to military historian Richard Saunders, "felt that the U.S. could not afford to stand alone in the face of global communist threats." After the crisis passed, "he devoted much attention to ensuring that no permanent damage to alliance relations was done."[75]

President Eisenhower's stance during the Suez conflict is understandable in terms of the conceptual framework presented in the opening chapter. There I asserted that the United States would not supply energy to its allies in support of strategic goals that conflicted with its own; instead it would exploit their dependence to achieve "coercive coordination." For their part, the allies thought that they could act unilaterally without fear of "abandonment" by Washington. The United States needed powerful allies for Free World defense, and the lesson they had drawn from Iran was that Washington would ultimately offer political and economic support.

But there were profound differences between Iran and Suez. In the latter case, the United States believed that a European victory would ultimately favor the Soviet Union, which had cast itself as Arab nationalism's friend and protector. According to foreign policy scholar John Campbell, Eisenhower was "unwilling to let the Soviet Union reap all the benefits of acting on behalf of the Arab peoples in a case like this in which the aggression was clear." By opposing the allies, he took "the chance to save some credit for the West. . . ."[76] Historian John Lewis Gaddis concurs, arguing that Eisenhower "won goodwill" in the Arab world after Suez. Richard Cottam corroborates the finding for Iran, stating that U.S. actions raised "American prestige . . . to a zenith."[77]

There were other differences as well. First, unlike the nationalization of Abadan, Nasser had stated clearly at the outset that he would pay fair compensation to Suez Canal shareholders. Second, Nasser quickly demonstrated his ability to operate the canal. Third, whereas the Shah represented an acceptable political alternative to Mossadegh, neither the British, French, Israelis, or Americans had found an alternative to Nasser. The three aggressors shared the hope that Nasser would be overthrown following their invasion, but by whom was unclear. Fourth, the Europeans had employed overt force against Egypt, in the absence of United Nations support and at variance with international law. Finally, the Soviet Union had threatened unilateral action against Britain and France, raising fears that the Suez conflict could lead to World War III.

But Eisenhower was not blameless for the alliance discord. At least to some observers it appeared that the United States was sending mixed messages to Europe regarding the use of force against Nasser. Chester Cooper interpreted such American proposals as the Suez Canal Users' Association as a delaying tactic, inspired by the domestic politics of a presidential campaign. The Dulles–Lloyd exchange quoted earlier suggests that the president may have been prepared to turn the other cheek had a quick and clean military operation been executed after Election Day. Even if Eisenhower was unalterably opposed to the use of force, he could have provided stronger support to the allies in their attempted negotiations with Nasser. He refused to consider the application of any economic sanctions after the nationalization, including the mildest forms like paying Suez Canal tolls to an escrow account rather than directly to the Egyptian government. Basically, Nasser never felt the consequences of his actions, despite the fact that Dulles considered them of dubious legality and a potential strategic threat to Western interests.

From an energy security perspective, the disruptions caused by the Suez Canal blockage were far more severe than those following the Abadan closure. But once again, the international oil companies demonstrated their flexibility in meeting supply shortfalls. Production was bolstered in the western hemisphere, and tankers were rerouted. The worst snags in the oil emergency plans were those created by the Texas Railroad Commission, highlighting the importance of domestic actors to policy execution.

In sum, at Suez the United States made clear that it would not provide oil in support of political objectives that conflicted with its own. This political fact presented ambitious European states with one option: to seek alternative suppliers who would prove as reliable as the United States in the event of a crisis. Indeed, as a reaction to American coercion, the Europeans would begin to develop their own power resources in the energy and military areas and in so doing contribute to the decline of hegemonic leadership. As historian Stephen Ambrose has concluded, "After the Suez Crisis the French, British, and Americans could never fully trust each other."[78] It was under this cloud of doubt that alliance energy security planning would have to proceed in the future.

6

Preparing for Next Time: 1957–1967

> Arab oil . . . is an Arab asset which can be put to use in the service of Arab aims.
>
> JOINT COMMUNIQUÉ, KHARTOUM SUMMIT, September 1, 1967[1]

The aftermath of the Suez crisis saw the allies agreed on the need for an energy diversification strategy, with the objective of denying Arab states future use of the oil weapon. This would be achieved by the development of new energy sources and suppliers, and the building of emergency oil stockpiles. The United States, however, did not want its allies to achieve complete energy independence from their leader. Instead, it sought to maintain the kind of economic leverage that could be exercised again, in the event the allies decided, as at Suez, to pursue unilateral foreign policies that Washington viewed as detrimental to its Cold War strategies.

For the European allies, however, energy security after 1956 meant diversification not only away from the Middle East, but from Texas as well. As Rand Corporation analyst Harold Lubell wrote of the Suez experience, "Even the United States turned out to be a politically unreliable source for Europe. . . ."[2] Thus, the allies began an elusive search for security without dependence.

During the Six Day War of June 1967, the Western alliance

would have the opportunity to test its post-Suez energy policies. Once again, the Suez Canal was blocked, forcing the allies to reroute their oil supplies. But 1967 brought the added dimension of a joint Arab oil embargo against Israel's Western supporters. While the west was able to overcome the energy shortage with relative ease, rifts in the alliance over Middle East policy indicated future opportunities for Arab leaders and the Soviet Union.

Nuclear Power and the Alliance

Among the alternative energy sources that received a post-Suez boost, none seemed more promising than nuclear power. The atom, it was commonly thought, would someday provide unlimited amounts of electricity at pennies the kilowatt. High hopes had been kindled ever since President Eisenhower's dramatic "Atoms for Peace" speech at the United Nations on December 8, 1953. On that day, he pledged U.S. support for peaceful development of nuclear power, so that the atom would "serve the needs rather than the fears of mankind."[3]

A number of international activities were launched following the Atoms for Peace proposal, most prominent being the establishment of an International Atomic Energy Agency. In Europe, the OEEC began to study the possibility of nuclear coordination under its auspices. This effort was largely supplanted by a June 1955 decision on the part of the European members of the coal and steel community to develop a supranational atomic energy organization, Euratom.[4]

An explicit set of U.S. objectives in the international nuclear area were spelled out in National Security Council statement 5507/2, which was prepared on March 12, 1955. According to the NSC, U.S. atomic technology was a "great asset in the effort to promote a peaceful world compatible with a free and dynamic American society." It could be used "to promote cohesion within the free world" and help assure the U.S. "access to foreign uranium and thorium supplies." At the same time, American-led nuclear programs would counteract Soviet advances in the atomic field.[5]

Accordingly, U.S. officials greeted the Euratom idea with enthusiasm. Secretary of State John Foster Dulles wrote Atomic Energy

Commission chairman Lewis Strauss that it would promote the creation of "a genuine United States of Europe." Given this political benefit, Dulles suggested that the U.S. "take active measures" to bring about "a Community of Six program in the field of atomic energy." But this meant the United States must be cautious in pursuing strictly bilateral agreements in the nuclear area, since these could conflict with the "larger" multilateral objectives of U.S. policy.[6]

On February 11, 1956, upon receipt of a report prepared by Louis Armand of France and revised by Paul-Henri Spaak of Belgium, the ECSC's Council of Ministers agreed to ask their governments to create an atomic energy community; West Germany abstained at this time owing to sharp internal divisions over Euratom. The community would have as its objectives the exchange of information in the nuclear energy field, the creation and formation of research installations, and the development of a set of safety regulations for peaceful nuclear activities. A tentative budget was suggested for Euratom, with allocations of $20 million during its first year of operation and $1.5 billion over the next five years.[7]

Despite vigorous support for Euratom on the part of such integrationists as Jean Monnet, initial debate over the terms of the treaty prevented rapid action within the member states. Then, on July 26, 1956, Colonel Nasser "made a decisive contribution."[8] The seizure of the Suez Canal highlighted Europe's growing energy dependence on the unstable Middle East. According to State Department official Gerard Smith, "The Suez adventure has resulted in a conviction that a closer association in Europe is indispensable. . . ."[9]

U.S.–Euratom talks during late 1956 and early 1957 were both spurred and complicated by Suez. On the one hand, it was clear that the allies needed diverse energy supplies; on the other, nuclear energy development would entail European dependence on Washington for enriched uranium. In November 1956, greater reliance on the United States was the last thing European officials sought.[10] From Gerard Smith's perspective, however, nuclear power relations held a key for renewing "Atlantic solidarity."[11]

The British government, which was not a party to Euratom, launched an urgent study of nuclear power during the Suez crisis, and in early 1957 it announced ambitious goals for power plant construction. Not wishing to be left behind, the Community of Six renewed their efforts at fashioning a treaty. A committee of "Wise

Men" was established, which visited Washington in February. John Foster Dulles gave the Wise Men a warm reception, arguing that nuclear power "may well open up a new area for European–US collaboration which can reestablish and strengthen the lines of cooperation between Western Europe and the United States which were strained by the Suez situation." He vowed that the United States would make available all the technical information, advice, and matériel that Euratom required.[12] On March 25, 1957, the Euratom Treaty was signed in Rome, along with a treaty creating the European Economic Community.

A U.S.–Euratom joint program was established soon after the treaty was signed, with the objective of helping Europe to achieve its goal of 15 million kilowatts of nuclear power in the six member states by 1967. Not only would the United States provide the technical support needed for reactor development, but it would provide substantial financial assistance as well. A reactor board of U.S. and European officials was convened in 1959 to consider proposals for power plant construction.[13]

Yet several factors contributed to undermine Euratom's mission from the outset. Within Europe, France remained committed to an independent nuclear program, especially after Suez. Humiliated by Washington, the French concluded, to quote the Gaullist journal *Carrefour*, that "the first lesson of Suez is that only possession of the atomic bomb confers power." Tensions quickly developed between Euratom and national nuclear programs.[14]

Equally significant in undermining the nuclear effort was the rapid change in international energy markets. After Suez, as we will see below, several new oil sources came on stream, dampening prices. At the same time, Europe became weighed down by a coal glut. Under these circumstances, nuclear economics looked increasingly unattractive. While European states were prepared to pay something of a "security premium" for diverse energy supplies, this had to be traded off against the economic welfare of their industries. As J. A. van den Heuvel of the OEEC's energy section said, "If the security argument alone is taken into account, there is a consequent risk of saddling the whole national economy with a heavy burden which may seriously impair its competitive position in the world."[15]

Thus, by the late 1950s nuclear power still remained an energy source for the future. Despite U.S. efforts to promote the technology within Europe, low-cost oil remained the continent's fuel of choice. It was in this market context that energy security policies were developed in the OEEC.

OEEC Energy Policy after Suez

The first energy security measure to receive attention from the OEEC in the aftermath of Suez was oil stockpiling. In 1958, the oil committee took up the idea of building a European oil stockpile. This continental stock would be owned by an international non-profit-making company with OEEC member governments acting as shareholders. In case of an emergency, the quantities owned by the European stockpile would be allocated by the Oil Committee of the OEEC to the member states.[16]

While the OEEC delegates decided not to execute this concept, they did agree to enhance national stocks. National stocks, the delegates felt, were necessary not only in the event of an alliance-wide shortage, but also in case of specific contingencies. Coal miner strikes, severe climatic conditions and oil interruptions were among the many occurrences that could lead public officials to release oil from a stockpile. National stocks provided governments with more flexibility than a Europe-wide consortium.

In July 1958 the oil committee suggested that each member have an emergency oil reserve of 90 days' consumption in place by 1960. This level would, according to the committee, enable member countries to survive a serious interruption of their supplies for a reasonable period of time. In fact, by 1960 the OEEC states had stocks varying in size from 61 to 79 days' consumption. The shortfalls were due to the difficulty of finding the funds with which to finance additional quantities.[17]

Regarding government–industry cooperation, the OEEC suggested that OPEG be prepared to meet again "at the first sign of emergency." This meant the creation of a "skeleton" organization composed of OEEC and oil company delegates that would be ready to work within 48 hours of a crisis. Further, the OEEC suggested that the U.S. government offer assurances that the MEEC or some equivalent grouping be reactivated during a shortage.[18]

In future, the OEEC offered the following emergency procedures. First, the chairman of the oil committee would ask the secretary-general of the OEEC to call a meeting of the organization's council to receive notification that a "state of crisis" existed. At the same time, the U.S. government would be asked to mobilize the MEEC. Within forty-eight hours, an "extraordinary session of the oil committee" would be held, attended by oil company representatives of OPEG and MEEC, with the objective of making a preliminary assessment of the shortage. If allocations were deemed necessary, the "Suez system" of distribution would again be adopted.[19]

The degree of confidence expressed by OEEC members in their ability to manage crises is noteworthy. The oil committee delegates were convinced that government–industry cooperation during the Suez crisis had quelled panic and prevented a scramble for oil. In 1958 the OEEC reiterated its belief that "in the event of a shortage of supplies oil should be shared on a fair and equitable basis." At the same time, it asked governments to provide "all necessary information relating to the oil supply situation," so that during a crisis every member state would have the same data concerning the market.[20]

Besides stockpiles and the formalization of emergency procedures, a major energy security initiative after Suez was diversification of oil supplies. By the early 1960s Western Europe was drawing oil from North Africa and the Soviet Union, in addition to the Middle East and western hemisphere. These new sources met nearly 19 percent of Europe's requirements in 1962. The perception that supply diversification provided enhanced security encouraged Europe to increase its oil consumption by 50 percent during the five-year period 1959–1964.[21] Ironically, this growth in oil utilization led to a shift away from indigenous coal; Table 6–1 illustrates the changing composition of Europe's energy supplies.[22] As will be seen in a later section, Europe's oil diversification strategies, particularly with respect to the Soviet Union, became yet another source of tension in alliance energy relations.

It should be emphasized that oil corporations no less than governments had an interest in diversification of oil concessions. As economist Helmut Frank has written, "Diversification could lessen dependence on the politically unstable Middle East and strengthen the companies' bargaining position vis-à-vis the governments of producing countries."[23] But this benefit had to be balanced against

TABLE 6–1. Western Europe's Primary Energy Consumption, 1957–1964
(in %)

Year	*Coal*	*Petroleum*	*Gas*	*Hydropower*	*Total*
1957	72	23	1	4	100
1959	64	30	2	4	100
1962	55	39	2	4	100
1964	49	45	2	4	100

Source: OECD, *Politique Energetique* (Paris: OECD, 1966), p. 31.

several problems posed by new supplies. First, the downward pressure on price would lower the value of existing concessions. Second, the ability of the companies to promote increased consumption of foreign oil would be hampered, as the following section shows, by autarkic U.S. energy policies. Finally, a new supplier—the Soviet Union—would enter markets with little respect for the existing rules of the game.

U.S. Energy Policy

Major energy security policies were also being developed in Washington after 1956. Prior to Suez, the United States imported 1.4 million barrels per day of petroleum, or 16 percent of domestic demand. Of these imports, somewhat less than 15 percent derived from the Middle East, amounting to an insignificant portion of overall consumption.[24] Despite the small volumes, domestic producers had become increasingly concerned by imports, especially since they sold at a lower price in East Coast markets. If import levels increased, high-cost domestic production must be shut down. The Middle East crisis played right into the hands of the Gulf Coast oilmen, providing them with a ready-made "national security" argument. This argument was used effectively over the following months and years to limit U.S. oil imports.[25]

In 1957 President Eisenhower appointed a Special Committee to Investigate Oil Imports, with the task of clarifying the national security arguments revolving around oil. After studying import patterns and the condition of the domestic industry, it concluded that "national security requires the maintenance of some reasonable balance between imports and domestic production." Specifically, the committee recommended that crude oil imports be "voluntarily" limited to 12 percent of domestic production. In July, the president accepted this voluntary approach.[26]

Unfortunately, voluntarism proved unworkable. The various oil importing companies were in such different positions regarding their oil supplies that they could not develop a formula that was acceptable to all. Some companies, for example, had substantial domestic production, so a small decrease in imports represented a tiny fraction of revenues. Firms that were more dependent on import earnings, in contrast, found the president's proposal burdensome. Over time, cheating became rampant, and by late 1958 the president had launched a new oil import study. This time, oil quotas were recommended.

On March 10, 1959, a presidential proclamation was issued establishing the Mandatory Oil Import Program (MOIP). This set the amount of imports allowed into the country and established a complex system to allocate import license—called "quota tickets"—to individual companies. Tickets were issued for both crude oil and refined products.[27]

President Eisenhower's proclamation was legally based on the national security provisions of the Trade Agreements Extension Act of 1954, which authorized the chief executive to control imports and exports when the nation's defense and security efforts were threatened. It is crucial to point out that the president justified his decision not simply on the basis of national security requirements narrowly defined but also in terms of the Western allies. Spare production capacity in the United States, the president argued, was vital to the alliance in the event of an emergency. Thus, while the MOIP was certainly encouraged by domestic political actors, it did have a broader strategic rationale in the eyes of the president.[28]

The impact of the MOIP was far-reaching. The quota policy severed the United States from the world market, permitting domestic oilmen to keep high-cost production in service. It "lowered the international demand for oil relative to its supply and put downward pressure on the international price for oil."[29] This outcome, in turn, had opposite effects in consumer and producer countries. America's Western allies welcomed the controls, since they resulted in both lower prices in Europe and the maintenance of U.S. spare production capacity, which was important in case of a future interruption of Middle East oil supplies. But producer countries, which lived off oil revenues, protested the decline in price, and in September 1960, representatives from Iran, Iraq, Kuwait, Saudi Arabia, and Venezuela met and agreed to establish an Organization of Petroleum Exporting Countries (OPEC) to coordinate oil policy decisions.[30]

Many recent analysts of the oil import program have been critical of President Eisenhower's decision, viewing it as a major blunder of U.S. energy policy. Political scientist Robert Keohane has argued that special interests effectively "drained America first" and "prevented the implementation of a farsighted strategic policy of conservation."[31] But it can be argued that a "free market" in petroleum would have "flooded America first." The resulting imports would have limited domestic production that could not have been turned on, like a water tap, in the event of a national emergency. Further, by removing the United States from the world oil trade, the allies benefited from lower petroleum prices; this served as a subsidy for their energy-intensive industries and contributed to their economic growth.[32]

This does not mean that a *quota* was the most efficient or equitable way to protect domestic oil producers. As economists Douglas Bohi and Milton Russell have argued, tariffs would have been equally effective yet less costly to consumers in meeting President Eisenhower's objectives. Quota tickets, however, gave the president the ability to allocate something of great value to oil importers, and this must have proved too politically tempting to resist.[33]

The United States also pursued several energy initiatives with more explicit alliance objectives. On May 13, 1959, the president approved a National Security Council action aimed at reducing European dependence on Middle East oil. The NSC proposed that

> in order to retard Western Europe's increasing dependence on Middle East oil and to reduce the effects on Western Europe of an emergency created by any complete or partial denial of Middle East oil resources,

> the United States should continue to encourage such action as is economically and politically feasible to facilitate the orderly development of alternative Free World sources of oil and other forms of energy outside the Middle East, and the broad diversification of means of transporting fuel in the Free World. The United States should also urge Western European countries to increase their petroleum stockpiles and to have in readiness emergency plans for conservation, sharing and transportation of oil.[34]

The NSC activated its proposals with three measures: first, geological studies were conducted of oil development possibilites in eleven countries outside the Persian Gulf; second, research and development of alternative energy sources was encouraged, with a focus on oil shale, tar sands, natural gas, and nuclear power; finally, the United States supported OEEC oil committee recommendations with regard to emergency planning.

In a June 1960 progress report on its energy policies, the NSC looked back with satisfaction at the progress that had been made over the previous year.[35] Although European oil consumption was climbing, a notable degree of diversification away from the Middle East had occurred. The only trouble spot in this strategy was the rapid growth in imports from the Communist bloc, a source that hardly seemed more reliable to Washington than the Arab sheikdoms. Table 6–2 illustrates the shifts in Europe's supplies between 1955 and 1960.

The Soviet oil dilemma will be discussed in the following section; here the growth in North African sources is described. The French discovered oil in the Saharan departments of Algeria, after intensive exploration, in 1956; Esso made its first discoveries in Libya in 1958. The U.S. government expected these new pools to have dramatic consequences for world energy markets. According to a State Department report, Persian Gulf producers would now "find themselves less able to use suspension of oil as a means of pressure." At the same time, the importance of North African oil and gas to France meant that it would be more determined than ever to win its battle against the Algerian revolutionaries.[36]

The introduction of new oil on world markets must place downward pressure on prices, which the State Department viewed as a problematic development. Petroleum products would become less

TABLE 6–2. Western Europe's Oil Supply and Demand Pattern: 1955 and 1960
(Thousands of barrels per day)

	1955		*1960*	
	No.	*%*	*No.*	*%*
Total demand	2,368		3,750	
Supply:				
Indigenous production	178	8	300	8
Imports				
Middle East	1,730	73	2,440	65
North Africa	—		180	5
West Africa	—		50	1
United States	50	2	30	1
Other western hemisphere	350	15	500	15
Communist bloc	60	2	250	7

Source: National Security Council, "Means of Retarding Western Europe's Increasing Reliance on Middle East Oil," June 23, 1960, *DDRS* (1981), doc. 338A.

expensive in Europe, "eliminating one of the competitive advantages of U.S. industries," especially since U.S. prices were now being kept artificially high by import controls. Lower prices also meant lower revenues for oil-producing countries, with potentially destabilizing results.[37]

Further, lower prices would have an impact on alternative fuels. Ever since the end of the postwar coal crisis, European governments had kept domestic coal industries alive with a variety of subsidies, including oil import quotas and taxes on fuel oil. After Iran and Suez, national security arguments were made by coal producers and miners in support of these measures. But a sharp decline in oil prices would raise the costs associated with such policies, and in light of international industrial competition, European manufacturers would seek energy price relief. The specter of a coal/oil price war emerged clearly on the horizon. Falling oil prices must also influence European perceptions of nuclear power, as discussed in an earlier section.[38]

From an energy security perspective, the National Security Council in 1960 perceived Western Europe's situation as improved but still in need of vigilant oversight and continued efforts at diversification and emergency management. As a result of the oil import program, the United States had "very large productive capacity and reserves" which "for the years immediately ahead provide Europe's principal safety factor in the event of denial of Middle East oil." These reserves would permit the United States to export 1.5 million barrels per day of oil at the outset of a crisis, rising to 2 million barrels per day after emergency plans were enacted. The NSC warned, however, that U.S. reserves were entering a period of decline, as domestic demand outpaced new additions.[39]

The most worrisome trend in Europe's energy outlook was growing reliance on Soviet oil. Further, as part of their "oil offensive," the Soviets had offered technical and financial assistance to the petroleum industries of such countries as Argentina, Brazil, Egypt, Iraq, and Syria. The United States countered Soviet efforts in specific cases; U.S. financial assistance was provided to the Bolivian government oil agency, YPFB. While the United States, in general, did not wish "to interfere in the flow of private capital," aid was used to prevent less developed countries from accepting Soviet assistance or "engaging in trade with the Bloc at levels sufficient to create undue economic dependence."[40]

The Alliance and Soviet Oil

In terms of alliance security, most troubling to Washington was the growing number of oil transactions between the Soviet Union and Western Europe that began after Suez. These oil sales were generally negotiated on a government-to-government basis, to the consternation of the international oil companies.[41] Sweden was the first Western European nation to engage in a major bilateral oil contract with Moscow, a direct result of the Suez crisis. Energy analyst Peter Odell has suggested that Sweden's neutral position on the continent and specifically its absence from NATO enabled it to strike this ground-breaking deal; by 1960, the Soviet Union supplied Sweden with 15 percent of its oil requirements.[42]

In the wake of the Swedish contract, several other countries followed suit. Austria signed a deal for oil in 1960 that resulted in its obtaining 22 percent of its oil requirements from Russia. In 1961, Italy obtained 14 percent, and West Germany and France somewhat under 10 percent, of oil needs from the Soviet Union. The Soviets sold their crude oil at such low prices that the international oil firms complained to their governments about an "oil offensive" being directed from Moscow. According to economist Helmut Frank, the price-cutting behavior of the Soviets forced the majors to follow suit, creating a downward spiral that caused the "breakdown" of a price structure already under pressure from new oil sources.[43] The companies were further troubled by the fact that many Soviet deals were of the barter variety, in which Western European countries exchanged steel pipe for oil.[44]

This "oil offensive" was disturbing to American policymakers as well. The State Department warned Western Europe of "the great economic risks involved in relying on an energy source which . . . could be turned off at the whim of the USSR. . . ."[45] The government cited Moscow's curtailment of exports to Israel following the Suez campaign as an example of Soviet economic warfare and made it clear to Europe that Russian oil was no more secure than Persian Gulf crude.

Nonetheless, increasing reliance on Soviet oil was one aspect of Western Europe's energy supply diversification strategy; European analysts reasoned that it was unlikely that *all* its oil suppliers would embargo the continent at the same time. The growth in the East–West oil trade during the late 1950s and early 1960s far surpassed predictions; between 1956 and 1962 OEEC consumption of Soviet oil grew by 700 percent, and several Western states came to rely on Moscow for more than 20 percent of their supplies. American policymakers were particularly anxious about Italy's 20 percent dependence on Soviet oil; the Italian Communist Party was the strongest Communist party in Europe, and policymakers feared that it would exploit East–West economic relations to its electoral benefit.[46]

Many East–West energy deals involved the exchange of European steel pipe for Soviet oil, and in the early 1960s the Soviets began building a new "Pipeline of Friendship" which would carry oil to the Baltic Sea, close to Western markets. "In 1962–63," Bruce Jentleson writes, "the Kennedy Administration moved to increase pressure on the West Europeans to stop selling the Soviets wide-diameter pipe and stop buying Soviet oil."[47] The administration was particularly determined to halt construction of the friendship pipeline.

In 1962, Secretary of State Dean Rusk circulated a telegram to American embassies in Western Europe that stated the administration's arguments against the pipeline. Rusk said it would

> facilitate and improve relative military, strategic and economic strength of USSR; (2) provide additional and less vulnerable means to supply petroleum to armed forces in Eastern Europe while permitting undetected build-up of petroleum stockpiles; (3) sharply reduce burden on overloaded transport, freeing facilities for carrying other critical logistical requirements.[48]

The American offensive against the pipeline incorporated several elements. First, in 1962 Congress passed the Export Control Act which permitted the United States to apply its laws to the foreign subsidiaries of American firms. Second, diplomatic pressure was placed on West Germany, the leading exporter of pipe. Third, the State Department lobbied the major oil companies to lower their prices in the European market. These actions delayed the completion of the pipeline by an estimated two to three years; further, the oil price reductions permitted Western firms to retain market share and limit the expansion of Soviet oil.

The pipeline episode is suggestive of the post-Suez changes in U.S.–European energy relations. While Washington retained sufficient influence to slow the pipeline deal, it could not stop the project's completion. Europe's oil diversification strategy meant that the allies were achieving greater flexibility in the energy area, something U.S. officials allegedly supported. American protests regarding Soviet oil use appeared contradictory, since the events of 1951 and 1956 had taught the alliance that Middle Eastern suppliers were unreliable and that alternatives were needed.

Conflict over Soviet energy imports had become an enduring theme in alliance energy relations. At the end of World War II, the

United States discouraged Western Europe from becoming dependent on Polish coal, fearing the economic leverage Moscow would gain as a result. Washington argued that Western Europe had ample indigenous resources, so coal policy should focus on mine rehabilitation rather than imports. But the same geological conditions did not hold for oil. The Europeans argued that they had no choice but to draw from the eastern hemisphere's great "puddles," including the Soviet Union. Questions concerning the prudent degree of dependence on Moscow were never resolved by the allies in the 1960s; indeed, they would reemerge in the 1980s as Western Europe assisted in the construction of a Siberian natural gas pipeline.

New Organizations: OPEC and OECD

The early 1960s saw the creation of two new organizations—the Organization of Petroleum Exporting Countries (OPEC), and the Organization for Economic Cooperation and Development (OECD)—both of which occupy a central place in the remainder of this study. OPEC was created in 1960 in response to falling crude oil prices on world markets and resultant declining revenues in producing countries. The organization's creation was surprising to many oil analysts at the time, in that producer country relations had been characterized by competition rather than cooperation. Cooperation, it was widely believed, would only occur in reaction to the establishment of a consumer organization (e.g., the European Oil Community, discussed in the last chapter).

Competition had ruled producer behavior during the postwar oil crises of 1951 and 1956. On both occasions, the major oil firms met supply shortages by boosting production in states that did not participate in the cutoffs. The oil producers were happy to compete with one another so long as higher prices meant greater revenues for economic development. But lower prices were anathema to all producers, and it was this situation that encouraged Venezuela's Minister of Mines Juan Perez Alfonzo and Saudi Arabia's Petroleum Minister Sheikh Abdullah Tariki to found OPEC.[49]

The Organization for Economic Cooperation and Development was created in 1961 as successor to the OEEC, and a major difference between the two was that the United States and Canada now entered the OECD as full members. Within a few years, Japan, New Zealand, and Australia would also join this grouping of industrial nations. According to a careful student of Atlantic economic relations, William Diebold, the replacement of the OEEC by the OECD reflected the new economic realities of the 1960s. The OEEC had been created to administer Marshall Plan aid, but by 1960 Europe was no longer relying on American assistance for reconstruction. The dollar and major European currencies were now convertible, and trade barriers between the Old and New Worlds had diminished. At the same time, European economic cohesion was growing in the European community, and the OEEC had become less of a focal point for economic policymaking. In Diebold's words, the creation of the OECD was something of a "rescue operation" in reaffirming the importance of Western economic cooperation.[50]

Yet Diebold states that the OECD agreement "seemed weaker than the old, and there was some fear that the new body might prove completely ineffectual."[51] Whereas the OEEC embodied rules and obligations that members were bound to follow, the OECD charter provided only for consultation and voluntary cooperation. The U.S. Congress refused to pass a treaty with stronger provisions.[52]

These changes were partially reflected in the energy area. A new OECD Energy Committee was established, responsible for the study of all energy-related problems. It was assisted by a Special Committee for Oil, which in effect was the old oil committee. At this time the European community was also launching energy policy studies, suggesting the continent's growing interest in formulating independent strategies without American input. The OECD brought with it both continuity and change in the energy issue area.[53]

The primary concern of the OECD oil committee was the West's reliance on OPEC oil. Imported oil met 20 percent of OECD–Europe's energy requirements in 1956 and 30 percent by 1960. While diversification of suppliers had been encouraged by governments to lower the risk of a cutoff, the creation of OPEC placed this strategy in doubt; the producers now presented a united front. In a 1961 report the oil committee stated:

> There must be doubts . . . on how far Western Europe could look to Venezuela to provide a substantial part of their emergency supplies—as it did in the Suez emergency—in the event of an interruption. . . . As a member, and a key advocate of OPEC, Venezuela might be disinclined to benefit itself at the expense of other members of the Organisation and, in any case, it is doubtful whether the Venezuelan Government would leave the oil industry as much freedom in increasing their output as they did on the earlier occasion. They would be likely to drive a hard bargain both over the quantities that could be exported and the prices to be paid.[54]

In the event of a solid OPEC embargo, OECD members would have no choice but to request emergency supplies from the United States.

The Six Day War

In 1966, OECD states consumed 1,106 million tons of oil. Western Europe relied on imports for 95 percent of its needs, and Japan, 100 percent; these areas imported over 80 percent of their petroleum from the Middle East and North Africa. Despite the MOIP, oil imports were also growing in the United States, rising to 21 percent of consumption in 1966.[55]

Western Europe's increased reliance on oil was dramatic. In 1960, oil supplied 30 percent of the continent's energy; by 1967 the figure was above 50 percent. The boom in demand was caused largely by the growth in automotive transport, the building of modern road networks, and overall economic expansion coupled with the fall in price and fuel-switching.

Nevertheless, Europe's energy security appeared to be stronger than ever. Oil stockpiles of 70 days' normal consumption were held by most countries (below the 90-day level the oil committee had recommended), and energy diversification was succeeding. Whereas in 1956 Europe obtained none of its oil from North Africa, in 1966 Algeria and Libya supplied 25 percent of imports. Further, as described previously, many countries were now receiving oil from the Soviet Union.[56]

In addition, the OECD had further institutionalized business–

government relations in the event of a supply emergency. The oil committee was empowered to establish an International Industry Advisory Board (IIAB) during an oil crisis, which would provide advice on the availability of oil and "assist in the implementation of the oil committee's recommendations for the apportionment of available supplies."

Yet creation of the OECD introduced a major chink in the energy security armor. For nearly a generation, the oil committee's emergency planning had focused solely on Western Europe. The OECD continued to maintain this Euro-centric system until a new one could be devised that accounted for the very different resource endowments of the United States, Canada, Australia, New Zealand, and Japan. In the meantime, these countries' requirements during an emergency would be met by ad hoc measures.

The various OECD energy security policies adopted after Suez appeared to be bolstered by the divisive politics that characterized the Arab world. Egyptian President Abdel Nasser had immersed himself in conflicts on the Arabian peninsula, and outside observers concluded that his preoccupation had shifted from destruction of Israel to defeat of the "reactionaries" King Hussein of Jordan and King Faisal of Saudi Arabia.[57] While he had created a defense pact with Syria, this was designed to increase his influence over that country rather than to prepare for a new offensive against the Jewish state.[58] Indeed, since the Suez war Nasser had been accused by more bellicose Arabs of hiding "behind the skirts" of the United Nations Emergency Force (UNEF) that was deployed along the Egyptian–Israeli armistice line.

Yet Nasser's cautious stance toward Israel was being overtaken by events in late 1966 and early 1967. A series of border clashes between Israel and Syria had heated up to the point where, by May, Israeli leaders were threatening military force against Damascus if terrorist activity did not cease. Nasser now faced a dilemma: he knew that the Arabs were too weak to defeat Israel, but he would lose face if he severed the defense pact with Syria. His initial response to the border problems was to urge caution on Damascus; at the same time he provided Syria with some military supplies. But the Syrians continued to adopt a bellicose stance, and on May 14 Nasser sent word to Damascus that, in the event of a war, "Egypt would enter the battle from the first minute."[59]

To back up his words, Nasser asked U.N. Secretary General U Thant to order the removal of UNEF. This force had been of tremendous value to Israel, especially its detachment at the tip of the Sinai peninsula which permitted Israeli ships free passage through the Gulf of Aqaba to the Red Sea. On May 22, despite diplomatic efforts on the part of the United States, Nasser declared that the gulf was part of "Egyptian territorial waters" from which Israeli vessels were excluded.

It was well understood by all observers that this action would not be tolerated by Israel. To the relief of the world community, both the United States and the Soviet Union sent letters to Middle East leaders urging restraint. The United States and Great Britain also drafted a "maritime declaration" stating Israel's right to free passage which they hoped all the European allies would sign; this hope was dashed by the intransigence of Italy and France.[60] For his part, French president Charles de Gaulle stated that if the declaration were to be meaningful, it must have the support of all the "major powers" rather than the Western allies alone. De Gaulle had made clear that he wished to distance France from American diplomacy in the region.[61]

Notwithstanding the flurry of diplomatic activity, it soon became apparent that the Middle East was heading toward another war. On May 30, the Arab world was treated to the sight of King Hussein in Cairo, where he signed a defense pact with his long-time adversary Abdel Nasser. This step, which proved to be a major blunder on Hussein's part, "convinced the Israelis that they no longer had any choice but to fight."[62]

On June 5, 1967, Israel launched a series of preemptive air strikes against its Arab neighbors. The surprise attack destroyed the bulk of Arab air power while it was still on the ground. With total control of the air, the Israeli army went on to control the Sinai peninsula and a chunk of Jordanian territory, including the Old City of Jerusalem. The Israelis then unleashed their fury on Syria; by June 10 they held the Golan Heights and were on the road to Damascus. That day hostilities officially ceased.[63]

On June 6, the major Arab oil producers, meeting in Baghdad, called a halt to all oil exports. Among the North African producers, Libya complied while Algeria continued to ship petroleum, with destination restrictions applying only to Great Britain and the

United States.[64] Use of the Iraq Petroleum Company pipeline and the TAPLINE, as well as the Suez Canal, was blocked. These oil arteries had been shipping nearly 5 million barrels of oil per day to OECD countries, or 65 percent of total supplies. For the first time the Arabs had joined together in brandishing the oil weapon.

This possibility had been foreseen in Washington and other alliance capitals. Indeed, even before the outbreak of hostilities, Iraq had threatened to embargo any Western state that supported Israel during wartime. In preparation, the White House formed in May a "Working Group on Economic Vulnerabilities" to assess the impact of an oil embargo.[65] This group presented its report on May 31, concluding that "the Arab countries together would have powerful economic weapons to use against the Atlantic nations. . . ." Not only could an embargo have a devastating economic impact, but Arab states might also expropriate the properties of U.S. and British oil companies, possibly selling them to French or Italian interests. With this fear in mind, the working group stated that Washington would have "to hold the Europeans to a common front. . . ." The United States must provide emergency oil supplies if requested and take appropriate domestic actions such as fuel rationing if this was necessary to meet an oil crisis.

American energy planning was complicated by the oil requirements of U.S. forces in Vietnam. These needs, amounting to over 200,000 barrels per day, were being met "almost entirely from Saudi Arabia and Bahrain." Obviously, if an embargo was launched, the United States would have to ship oil directly to Vietnam, making it that much more difficult to supply Europe and Japan. Based on these findings, White House adviser Walt Rostow suggested that Washington consult immediately with its NATO allies, especially Britain, on the related problems of oil and finance. Rostow met with British Ambassador Patrick Dean on June 1 and with the other NATO ambassadors on June 5.[66]

A variety of U.S. government assets, including the CIA, were mobilized to study alternatives for overcoming an alliance oil shortage. In a report prepared on June 7, the day after the embargo was announced, the CIA projected Western European and Japanese requirements and alternative suppliers. The agency estimated that stockpile drawdowns would be necessary, while non-Arab oil producers, "by the intense use of existing facilities," could meet the

TABLE 6–3. Emergency Energy Supplies for Western Europe and Japan: 1967
(Thousands of Barrels per Day)

	Western Europe	*Japan*	*Total*
Normal supply			
Arab oil: Middle East	3,800	1,160	4,960
Arab oil: North Africa	2,100	0	2,100
Iran	800	500	1,300
Other	1,600	340	1,940
Total	8,300	2,000	10,300
Emergency supply (first six months)			
Iran (normal supply)	800	500	1,300
Other (normal supply)	1,600	340	1,940
Increase in (Iran)+(Other)	1,480–1,710	280–320	1,760–230
Withdrawal from stocks	2,770	670	3,440
Total	6,650–6,880	1,790–1,830	8,440–8,710
Percent of normal	80–83	90–92	82–85

Source: CIA, "Impact on Western Europe and Japan of a Denial of Arab Oil," June 7, 1967, *DDRS* (1981), doc. 280A.

bulk of OECD needs. But it would take time for these new supplies to come on stream; in the near term, Europe and Japan faced shortfalls of up to 20 percent.[67] Table 6–3 provides the CIA projections of the energy shortage.

From an economic standpoint, the CIA was particularly concerned about the oil shortage's impact on Britain's balance of payments. The agency projected that the shift from sterling to dollar oil would raise Britain's dollar expenses by $300 million per year, causing severe economic problems for America's closest ally. The United States feared that these economic problems would translate into lower readiness on the part of Britain's NATO components.

On the basis of the various economic studies performed after the outbreak of the Six Day War, Assistant Secretary of State for Economic Affairs Anthony Solomon outlined a plan of action on

June 9. Solomon made clear that the Europeans and Japanese would place "political pressures" on the United States to allocate oil if they faced continuing shortfalls. It was also possible that the Arabs would pursue side deals with the Western allies. Solomon reminded his administration colleagues that "the Europeans and Japanese have long wanted to get direct access to Middle East oil sources" and might be tempted to displace Anglo–American companies if the opportunity arose.[68]

Accordingly, Solomon recommended that the U.S. cooperate with the allies, stating that "the immediate focus for managing the oil shortage and tanker availability is the OECD." A cooperative stance required in the first instance that the secretary of the interior declare an emergency, which would permit American oil companies to participate in an OECD-sponsored allocation scheme. Further, Washington should discuss with its allies "the possibility of increasing production in the U.S. This in itself would be reassuring to the Europeans and Japanese, and provide some deterrent against immediate temptation to make side deals." The United States should also prepare rationing plans, which would symbolize America's willingness to share the burden of an oil crisis.[69]

On June 10, the secretary of the interior proclaimed the existence of an oil emergency, and once again the Foreign Petroleum Supply Committee was activated. As in 1956, this committee, composed of oil company executives, was empowered to discuss means for overcoming any supply shortages. Actual implementation of plans, however, would require an OECD oil committee motion stating that an emergency or the threat of an emergency existed; otherwise, the U.S. Justice Department would not waive antitrust laws that prevented the firms from sharing market information.[70]

Such a motion, requiring oil committee unanimity, was not forthcoming when the OECD delegates met on June 12. The day before, Saudi Arabia and Abu Dhabi had resumed oil exports to all destinations except the United States and Britain, and it appeared that the other producers would soon follow suit. Stock levels in OECD–Europe and Japan were high, so that drawdowns could occur if necessary. Despite the continuing closure of the IPC pipeline and TAPLINE, and the Suez Canal, many OECD members were confident that they could obtain sufficient supplies and that it would not be necessary to declare an emergency and activate allocation procedures.

The United States delegation reacted with disbelief to the oil committee's inaction. The member states were forcefully reminded that the United States had initiated the internal legal procedures that were required before an emergency could be declared; now it was being greeted by a complacent group of allies. The U.S. delegate warned that American-based oil companies would not be authorized to share information with their foreign counterparts in the absence of a declaration that an emergency or the threat of an emergency existed. On the basis of this speech, the OECD members were driven to reconsider their views, and the oil committee finally adopted a motion—with France, Germany, and Turkey abstaining—that the "threat of an emergency" existed.[71]

Alliance relations within the oil committee suggest that differing Middle East strategies may have complicated emergency planning. President De Gaulle had worked hard to distance his government from its traditional pro-Israel stance, asserting that Paris opposed the side that had initiated military action. De Gaulle's position was welcomed in the Arab world (if not in France, where public opinion remained pro-Israel); the Syrian Foreign Minister expressed appreciation for a French policy that "differed conspicuously from the aggressive attitude of the United States."[72] To one astute observer it appeared that the general had "imposed upon his country a policy much closer to that of the Soviet Union than to that of any allied government."[73] Historian Walter Laquer has characterized the alliance response to the Six Day War as one of "disarray."[74]

Nonetheless, the interdependent nature of the oil system meant that individual OECD countries would find it difficult to obtain adequate supplies at the expense of their fellows. A state would have to possess not only a bilateral contract for oil from a producer, but sufficient tanker capacity to cart the oil and the appropriate refineries to make products. The only OECD member state vertically integrated in this fashion was the United States. The OECD members may have developed alternative energy suppliers after 1956, but they still needed the United States in an emergency.

Although the oil committee had determined that a formal allocation scheme would probably not be required to overcome the war-induced shortfalls, the closure of the Suez Canal and TAPLINE created massive logistical problems for the international oil companies. The major European and American firms, joined in the OECD as the International Industry Advisory Board, drew up plans that called for increased production in Texas and new tanker schedules. North African sources became especially valuable, given the short haul across the Mediterranean. During the time it took to execute these plans, Europe began to use some of the oil held in the stockpiles that had been created after the Suez crisis of a decade ago.[75]

From mid-June onward the oil supply situation appeared increasingly favorable. Kuwait, Bahrain, and Qatar commenced exports on June 14, and on June 23 Iraq reopened the IPC pipeline. On July 4, Libya began shipping oil to France, Greece, Italy, Spain, and Turkey. The Arab oil embargo was breaking down.

But these gains were partially offset by Nigeria's sudden disappearance from world oil markets. On May 30, the Eastern Region of Nigeria seceded, proclaiming itself "Biafra." In July the federal government ordered the blockade of the Bonny Oil Terminal, bringing exports to a halt. Nigeria had been the world's tenth largest producer, with exports of 384,000 barrels of crude per day. Forty percent of this oil was shipped to Great Britain, 15 percent to West Germany, and 10 percent to France.[76] Suddenly, the OECD countries faced a "dual crisis."

In meeting the Nigerian and Middle East shortages, the United States played a decisive role, as illustrated in Table 6–4. By July, shipments of crude oil from Gulf Coast ports were running at 650,000 barrels per day above normal. This time, with Texan Lyndon Johnson in the White House, the TRC lost no time in raising its allowables. Venezuelan crude production was also pumped up by 500,000 barrels per day, shattering the myth of OPEC solidarity. Tankers that had been mothballed or were plying the grain or mineral trades returned to oil service, and government-owned ships were released for emergency runs to Europe. By early August, the Foreign Petroleum Supply Committee could report, "There are sufficient supplies of crude oil and products to meet 3rd Quarter 1967 consumption requirements."[77]

Despite the allies' success in meeting the oil shortage, many Arab states were determined to maintain the embargo against the United States and Britain. A contentious meeting of Arab states in mid-August placed these hard-liners, led by Iraq, against Saudi

TABLE 6–4. U.S. Oil Exports, 1966–1968
(Thousands of Barrels)

	1966	*1967*	*1968*
Crude oil exports	1,477	26,541	1,802
Product exports	70,923	85,519	70,923

Source: American Petroleum Institute, *Petroleum Facts and Figures* (Washington: American Petroleum Institute, 1971).

Arabia, Kuwait, and Libya, which had gained nothing from the war and which stood to lose an important role in world oil markets. The embargo issue was only resolved at an Arab summit meeting in Khartoum at the end of the month, where it was decided to abandon all sanctions against the West and permit producers to resume shipments. The three early dissenters paid for this decision by agreeing to provide Egypt, Jordan, and Syria with an annual subsidy of nearly $400 million.[78]

Because of spot shortages resulting from the Biafra war and the closure of the Suez Canal, the IIAB continued to meet on a regular basis throughout 1967. By early 1968 the oil committee was convinced that the supply situation was "well in hand." Although the committee did not formally disband the IIAB at this time, it was demoted to a standby basis.[79]

The allies had thus managed another postwar energy shortage without divisive scrambles for fuel or side deals with producers. Owing to strong U.S. leadership, close business–government cooperation, and the variety of energy security measures enacted after Suez, the oil crisis of 1967 was overcome. The Arab oil embargo was decisively broken, as the United States bolstered production and other OPEC members took advantage of their brethren to gain market share. Indeed, some observers suggested that the events of 1967 spelled "the end of OPEC as a real force in the world of oil."[80]

Conclusion

The 1967 oil crisis marked a watershed in alliance energy relations. For ten years, the allies had been preparing for the next shortfall.

Stocks had been built, supplies diversified, and reserve levels in the United States maintained by oil import controls. A powerful energy security arsenal had been built.

But the mobilization of this arsenal required alliance policy coordination. In essence, such coordination could only take place on Washington's terms. As the U.S. delegation to the OECD made clear, the United States would not authorize the execution of its energy security procedures in the absence of an emergency declaration.

Initially, it seemed that such a declaration would not be forthcoming. The Arabs had employed a selective embargo, and the OECD "haves" were not necessarily interested in associating with the "have-nots." High stock levels and the success of their energy diversification strategies had emboldened certain OECD states to adopt Middle East policies that differed from those of the alliance leader. The leader was confronted with the ironic fruits of earlier efforts in the energy security area.

But in 1967 the United States still held a dominating position in world oil markets. It was able to increase indigenous production significantly to meet alliance requirements and to mobilize the tanker fleet needed for shipment. Further, no other Western power could challenge the Soviet Union in the Middle East. This latter fact was evident to allies and moderate Arab regimes alike.

During the Six Day War, alliance relations were characterized by a mixture of conflict, tacit coercion, and ultimate policy coordination. Taking a page from the Israelis' book, the Johnson administration had launched a "preemptive" strike on the oil embargo, offering the allies solutions before they even knew what the problems were. Indeed, a White House economic group had prepared for an embargo in May, one month before the war began. While the 1967 episode offers another example of "coercive coordination," with the United States refusing to share oil until the OECD found that a "threat" existed, alliance conflict was much less sharply drawn than at Suez.

Gazing over the broader panorama presented by the post-Suez decade, the period was characterized by "subsurface change." During these years OPEC and the OECD were created, Europe began to import oil from the Soviet Union, the United States enacted oil import quotas, and the allies began to distance themselves from support of Israel in the Middle East conflict. Each of these changes marked a subtle shift in the alliance energy security structure, but none of them appeared to undermine it. As a result, the allies would enter the 1970s confident that security could be maintained while oil imports were increased.

7

Allies in Conflict: 1968–1974

Oil, it has been said, seems to bring out the worst in nations.
WALTER LAQUER[1]

The Arab oil embargo of 1973–1974 shattered the energy security structure built by the Western allies after Suez. That structure was already showing strains during the Six Day War, as some of the allies sought to dissociate themselves from U.S. policies in the Middle East. But the underpinning of the structure had remained in place, as Washington still served as emergency supplier of last resort. By 1973 America's spare production capacity had vanished, and the U.S. proved unwilling to overcome the oil shortfall. The absence of hegemonic resources, on the one hand, and conflicting western interests in the Middle East, on the other, combined to produce the gravest alliance crisis since 1956.

And yet, by early 1974, the allies were attempting to rebuild their collective energy security with the establishment of the International Energy Agency (IEA). In proposing the IEA, the United States had explicitly linked alliance energy and military security, threatening uncooperative allies with a partial NATO troop withdrawal. This raised the costs of defection from U.S. leadership, causing the allies to reassess their foreign and energy policies.

The present chapter does not attempt to provide a comprehensive account of the 1973 energy crisis and its aftermath. Instead, the focus remains on alliance relations. The case of the oil embargo offers powerful support for the thesis that alliance cooperation has been most effective when hegemonic power was exercised in support of a common strategic purpose. In the absence of either power or collective interests during a crisis, alliance discord has been the result.

The Oil Market: 1968–1973

In focusing on alliance relations, this chapter pays more attention to the demand than to the supply side of the oil trade. But in order to place the oil crisis of 1973 into perspective, it is necessary to examine some of the changes in world oil markets that occurred after the Six Day War. Since many works provide extended treatment of this topic, this discussion will be brief.[2]

Following the Six Day War, the oil market began to tighten. The Suez Canal was closed during the war (it remained so until 1975), and Nigeria was wracked by civil war. Production in Iraq became unreliable as a series of disputes between the government and the Iraq Petroleum Company (IPC) disrupted exports. The demand/supply balance was further shifted toward the producers in May 1970, when an accident halted oil shipments through the Trans-Arabian Pipeline, which had been carrying 500,000 barrels per day to its Mediterranean outlet.[3]

A most interested observer of these changes was the new Libyan leader, Muammar Qadaffi, who had led the overthrow of King Idris in September 1969. Libyan oil had several advantages in the developing marketplace. First, Libya was only a short hop across the Mediterranean from several European refineries, lowering transportation costs. Second, its crude was "sweet" (meaning it had low sulfur content), making the fuel preferable from an environmental standpoint. Qadaffi recognized that these quality and location differentials could be exploited to obtain higher prices for Libyan crude (see Figure 7–1).

Colonel Qadaffi had an additional advantage in bargaining for higher prices—the structure of the Libyan oil industry. Unlike most Persian Gulf producers, who were dominated by one or two of the "seven sisters," the major producer in Libya was Armand

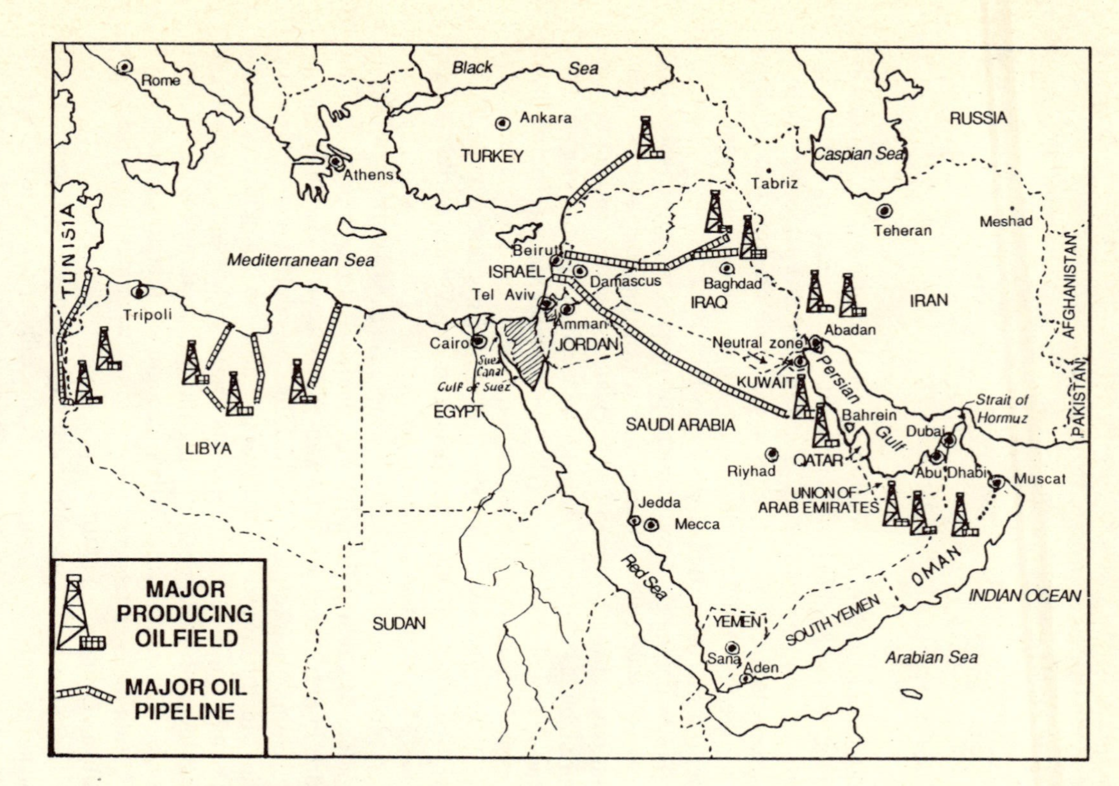

FIGURE 7–1. Middle East Oil Fields, 1970

Hammer's Occidental Petroleum Company. Oxy relied on Libya to provide 100 percent of its eastern hemisphere needs, and thus it lacked the flexibility of the big multinational firms. In September 1970, Hammer agreed to pay Libya an additional 30 cents per barrel for its crude, and to an increase in the tax rate on local company profits from 50 to 58 percent.

Libya's success was not lost on other oil producers. At the December 1970 meeting of OPEC, the organization called for a round of "regional" oil negotiations, in which the oil companies would hold separate price discussions with producers in each region. Naturally, the oil companies feared that such talks would lead to "leapfrogging" and higher prices; they sought a global agreement. OPEC refused, and in February 1971 the regional talks began in Tehran as the companies negotiated an agreement with the Persian Gulf producers. These led to substantial price increases ($.33 per barrel immediately, followed by $0.05 per year out to 1975) in return for "security of supplies" and "financial stability." In the Mediterranean, regional oil talks were subsequently held in Tripoli, where in March Libya won even higher prices than those negotiated at Tehran. As the oil companies had feared, the Persian Gulf producers protested, demanding Libya-like terms. A huge transfer of wealth from consumers to producers was under way. The Tehran agreement alone resulted in an additional $1.2 billion in royalty payments in 1971.[4]

Ironically, the State Department had sent Under Secretary John Irwin to Tehran in January to win the Shah's support for global talks. Once there, however, the Shah convinced him that such talks would be destroyed by OPEC "extremists"—i.e., Libya. The Shah argued that regional talks were in everyone's long-run interests, since they would better stabilize producer–consumer relations. Irwin accepted the Shah's reasoning and ended up pressuring the companies to accept regional negotiations. The "Irwin mission" would live in infamy in the hearts and minds of oilmen.[5]

The producers were also beginning to assert their control over industry operations. In December 1971 Libya nationalized British Petroleum's local concession, and the following June Iraq nationalized an IPC concession. In the wake of these events, the oil companies accepted a general agreement on participation with the Persian Gulf producers, which provided for local governments to obtain a 25 percent share in their oil operations no later than January 1, 1973, with increases ultimately leading to 51 percent government ownership no later than 1982. This schedule was quickly accelerated in Venezuela, Iran, and Libya, all of which either nationalized or assumed effective control of oil company operations by 1973.[6]

On the demand side, the decisive change in the market after 1967 was the rapid growth in U.S. oil imports. During the 1960s, U.S. spare production capacity had greatly diminished, as no major discoveries were made in the lower 48 states. Whereas in 1965 new discoveries added 3 billion barrels of oil to proved reserves, by 1969 the number fell to just over 2 billion. This meant that Americans were replacing the domestic oil they consumed in decreasing increments. Further, development of Alaska's North Slope and the Outer Continental Shelf was blocked for environmental reasons. Other environmental regulations encouraged the use of petroleum by electric utilities in order to cut down on coal-borne sulfur emissions. As a result, in 1970 the United States imported 165 million tons of petroleum, up from a 1967 level of 95 million tons. This figure was projected to rise to over 200 million tons by 1975 even if North Slope oil came on stream, or 24 percent of U.S. demand. This sudden boost in imports, a U.S. government official told the OECD oil committee in 1970, could lead to a period of tight supplies and higher prices.[7]

Owing to the growing demand for oil by the American economy, quantitative restrictions on imports were gradually relaxed.[8] This, as could be expected, caused domestic producers to rail, and in 1970 a cabinet task force was formed to study the "oil import question." The task force argued, to the producers' chagrin, that "national security will be adequately protected by adopting as a first step a revised control system and a . . . reduction in import restraints."[9] The protectionist system emplaced after Suez was being dismantled.

The report's comments on alliance relations are especially pertinent. The task force stated that, in the event of a Persian Gulf conflict or embargo, the United States would be incapable of supplying "anything like the deficit in . . . free world requirements brought about by a prolonged curtailment of Eastern Hemisphere supplies." It suggested that U.S. policy encourage the Western allies to "develop their own emergency resources."[10] This comment marked a profound change in alliance energy relations. At the end of World War II, the United States had developed an oil policy based on hemispheric self-sufficiency. Energy security in that context was provided by conservation of U.S. reserves, which were available to meet alliance emergencies. The policy had endured for twenty-five years; indeed, it was a cornerstone of alliance economic relations. Now, the United States could no longer support its energy commitments.

But President Nixon, recognizing the foreign policy implications, refused to abandon the oil import quota system. The president argued that such a decision "must necessarily await the outcome of discussions with Canada, Mexico, Venezuela and other allies and affected nations. . . ."[11] On the basis of these talks, Nixon maintained the quotas, only permitting additional imports of 100,000 barrels of oil per day. Clearly, the president did not wish to undermine the oil security pillar of the Western alliance.

But another presidential decision had that very effect. On August 15, 1971, Nixon ordered a general wage and price freeze in order to harness growing domestic inflation; petroleum prices were included in the freeze. This was the first in a series of energy price controls that would endure up to the Reagan administration. Reviewing this history, economist Joseph Kalt has concluded that price controls stimulated U.S. demand for oil and could be considered "consistent with foreign policy interests in amicable relations with certain exporting countries."[12] Since Middle East oil was much cheaper to produce than domestic crude, price controls favored foreign suppliers and discouraged domestic exploration and development. Perhaps more than any other policy decision, price controls would prove disastrous to U.S. and alliance energy security.

Oil price controls seriously exacerbated America's growing energy problems. Domestic oil reserves declined, and demand for foreign oil increased. In April 1973 Nixon had no choice but to suspend the mandatory oil import program. The domestic oil industry was now being slammed by price controls on one side and by cheap oil imports on the other. America's hegemonic power in

TABLE 7–1. U.S. Oil Statistics, 1970–1974
(millions of barrels)

Year	*Consumption*	*Imports*	*Proved Reserves*
1970	5,365	1,248	39,001
1971	5,553	1,433	38,063
1972	5,990	1,735	36,339
1973	6,317	2,284	35,300
1974	6,078	2,231	34,250

Source: Craufurd Goodwin, *Energy Policy in Perspective*, pp. 693–694.

energy was coming to an end. Table 7–1 provides some relevant indicators.[13]

The Alliance Response

A dramatic shift in alliance attitudes toward the emerging energy market can be detected between the late 1960s and early 1970s. In 1969, despite concerns over dependence on the Middle East and the decline in U.S. spare capacity, the OECD oil committee took a fairly optimistic view of energy security. Western Europe was succeeding in its energy diversification plans and now drew half of its petroleum supplies from sources outside the Persian Gulf; at the time of the Suez crisis that region supplied nearly 85 percent of the continent's oil requirements. Further, the experience of the Six Day War suggested the flexibility of the international oil industry in overcoming temporary disruptions. A final security factor was the presence of stockpiles averaging 70 days' consumption in OECD member states. Overall, it appeared that, while continued vigilance was required, there were no immediate threats to alliance energy security.[14]

Within a year the atmosphere had changed considerably. Libya's aggressive negotiations with Occidental, coupled with the Tehran–Tripoli agreements, suggested that the balance of power in oil was shifting away from consumer countries. The OECD estimated that Western Europe alone faced increased oil costs of $500 million per year owing to the regional agreements, and prices were only headed skyward. The American energy glutton had refused to limit its appetite for foreign oil, creating further pressure on prices. In addition, the absence of spare capacity made it increasingly likely that an oil embargo could be successfully executed since the United States would be unable to serve as marginal supplier. In light of the rapid changes in oil markets, the oil committee sought to develop an appropriate set of policies.[15]

It was clear to the delegates that national stockpiles had a special role to play in the new environment. On June 15, 1971, the committee recommended that states achieve a stock level of at least 90 days' average consumption. It noted that the possibility of a supply disruption remained and that the western hemisphere was unlikely to overcome any shortages that resulted. National responses to energy crises were coming to the fore.[16]

In most OECD countries, stockpiles were maintained not by governments but by the oil companies. While oil companies generally maintained 30 days of crude oil on hand to ensure smooth refinery operations, official pressure was placed on the firms to raise these levels to 90 days. Thus, much of the financial burden associated with this first line of defense was borne by private enterprise, though in some cases governments allowed firms to recuperate these costs through higher product prices.

Despite the growing emphasis on national measures, the U.S. delegation continued to promote alliance cooperation in the energy area. In 1972 a U.S. official stated,

> It is imperative for the world's major consumers of oil and other forms of energy to take joint and co-ordinated action—starting now—to increase the availability of all types of energy resources; to lessen, to the degree possible, an overdependence on oil from the Middle East, to co-ordinate the response of consuming countries to restrictions on the supply of Middle East petroleum, and to develop jointly and cooperatively a responsible programme of action to meet the possibility of critical energy shortages by the end of this decade.[17]

Accordingly, the oil committee continued to refine its emergency allocation system. In February 1972, the delegates agreed to a new system based on a "two-tier" approach. The first tier of oil supplies would be composed of 90 percent of the amount of crude and petroleum products that could be made available to OECD–Europe during a crisis. This oil would be allocated in proportion to each member state's "normal consumption." The remaining 10 percent would be available for special allocations to countries in dire need, as the oil committee decided. The provision of this additional amount would depend on the seriousness of a country's economic difficulties, climatic and seasonal factors, and the availability of energy alternatives.[18]

It is critical to remember that the oil committee's allocation system continued to apply only to OECD–Europe; other member states, including Japan, the United States, and Canada, would be incorporated on an ad hoc basis. This caused the Japanese particular anxiety, since many policymakers in Tokyo believed that, in a future crisis, "governments would direct the international oil companies to give their home countries preferential treatment."[19]

In response, four new Japanese oil companies were created in early 1973, charged with developing concessions in Asia and the Middle East. The Japanese government also engaged in direct negotiations with producer states, and in April the Minister of International Trade and Industry, Yasuhiro Nakasone, visited the Persian Gulf. Nakasone offered Japan's help in building new industries in Saudi Arabia, Kuwait, and Abu Dhabi. Further, he pledged that Tokyo would not participate in any "common front of oil consumers" directed against OPEC.[20] By 1973, 10 percent of Japanese oil supplies were imported on the basis of government-to-government agreements.[21]

The Japanese were not alone in bilateral dealing. France, Italy, Britain, and West Germany were among the other OECD members that sought supplies on a direct government-to-government basis during 1972 and 1973. The OECD oil committee noted with concern the growth in bilateral deals between producer and consumer governments which "carry with them a real danger of launching all consumers into a competitive cycle. . . ." Bilateral deals sabotaged the position of the oil companies, and the committee bemoaned the "absence of acceptable industry–government communications." These new producer–consumer relations threatened a fundamental principle of alliance energy security: that states refrain from "scrambling" for supplies. But the structure that had long preserved alliance energy security was beginning to fray by the early 1970s, rattled by changes on both the demand and supply sides of the oil equation.[22]

The oil companies were also dismayed by these bilateral activities, which undermined their traditional market position. In May 1973 the group planning division of Royal Dutch/Shell released a paper stating that an "oil scramble" was being unleashed by the industrial states. The bilateral deals were "designed to procure increased national security for energy supplies" but instead were causing a "rapid escalation of costs" for consumers. The paper argued that market chaos must result from this process.[23]

Shell believed that the "best way of reducing the sense of insecurity" lay in the "promotion and establishment of institutions with the authority to allocate oil resources." The OECD oil committee could conceivably do the job if its powers were "enlarged and widened." A deterrent must be developed to prevent participants from defecting from allocation agreements, though Shell admitted that these would not be easy to find; the most effective deterrent must be a nation's desire "to avoid public censure." Further, Shell recommended that the OECD–Europe allocation system be extended to incorporate other major consumers.

The oil firm noted that an alternative approach was being studied by Japan's MITI. This involved the creation of an "international energy agency" which would be composed of OECD nations, oil producers, and Third World states. The agency would not only have allocation responsibilities, but also funds for energy research and development. Shell expressed skepticism over this grandiose proposal, and doubted its ability to "help in mitigating the present potential scramble."

Shell recognized that its call for stronger government action in the international oil field was "paradoxical." But "times and circumstances change," the company said, and the only alternative to business–government cooperation was oil market anarchy. Clearly, the company feared for its position in the event of a future crisis. The opposing pressures from producer and consumer governments might tear the oil industry apart.

For its part, the United States continued to make new proposals to the oil committee to strengthen existing procedures. In July 1973, an American delegate, Robert Ebel, presented a paper titled "Alternative Criteria for Sharing Oil Supplies." Ebel suggested that the OECD must consider among three allocation alternatives: (1) no sharing; (2) sharing on the basis of preshortage *consumption* levels; (3) sharing on the basis of preshortage *import* levels. He stated that the first had "little to offer," since it must lead to competitive bidding for available supplies. The second would take oil imports away from the United States and move them toward Europe and Japan, since America's indigenous production was so great. The third would simply allocate the amount of available foreign oil to all consumers on a precrisis basis. Ebel felt that some compromise position between the second and third alternatives must be reached. Any scheme, he said, must be "OECD-wide" if it was to be effective.

The failure of the OECD to develop a new allocation scheme, combined with counterproductive U.S. energy policies, greatly weakened the allies' confidence in their energy security after 1970. Accordingly, they devoted increasing attention to their diplomatic relations with Arab states. Bilateral arrangements, distance from Israel, and national energy stockpiles appeared to provide the only chance for security in an increasingly unstable environment. From the allies' perspective, the United States had become a threat to, rather than a guarantor of, Western energy security.

War and Embargo

Exacerbating the strain in international energy markets was the threat of another Arab–Israeli war. The Six Day War had produced a brilliant Israeli victory, leaving the Arab belligerents humiliated. Israel had expanded into the Sinai and the West Bank of the Jordan. It held the Golan Heights and Jerusalem. The Jewish state was stronger than at any point in its history, the dominant Middle East power.

But strength produced smugness rather than wisdom. The Israelis stated that nothing less than direct negotiations with the Arab states would serve as a path to peace, thus posing conditions that they knew were unacceptable. For their part, Arab extremists continued to support terrorism and call for Israel's extermination, while more moderate leaders were frozen by anxiety over their own positions. Nor did the United States and the Soviet Union calm Middle East tensions. Rather than work jointly to solve regional problems, the superpowers left détente at the European border to pursue their rivalry abroad.[29]

In 1969, fighting broke out between Israel and Egypt along the Suez Canal, marking the outset of the "war of attrition" which was designed to sap Israel's stamina. A number of dramatic terrorist attacks took place both inside and outside Israel, with commercial aircraft becoming a growing target of Palestinian commandos. In 1970 and 1971 the United Nations and the United States made several attempts to halt the violence through indirect talks with the belligerents. The Israelis refused to participate in such discussions until the Egyptians removed the Soviet-installed missile batteries along the Suez Canal, which had been emplaced in violation of earlier U.N.-sponsored agreements.[30]

While these talks were unsuccessful, the chance of an all-out war appeared to diminish in 1972 when Egypt's President Sadat demanded the withdrawal of 20,000 Soviet advisers based in his country. Nonetheless, Soviet military supplies continued to pour into Egypt, and a number of advisers remained in place. These shipments were matched by American arms transfers to Israel, including advanced Phantom fighter–bombers.

By the spring of 1973, Sadat had made his decision for war. Not only was his military ready, but an unprecedented degree of Arab solidarity had formed around the issue of liberation of the occupied territories.[31] At the same time, Sadat believed that international politics now favored the Arab states. The Europeans, concerned for their energy security, would refuse to support Israel in wartime. Further, they would pressure Washington to support the Arabs so as to avoid another oil embargo, as the producers had launched during earlier Arab–Israeli wars.[32]

Indeed, winning producer state—especially Saudi—support for Egypt's military quest was an important element in Sadat's strategy. Like Nasser before him, Sadat believed in Arab oil power and the enormous leverage that this power could exert in world politics. He was well aware of the turmoil in the energy markets and of America's growing imports. He also knew that the United States was no longer in a position to meet its allies' needs in the event of a major oil crisis. In August 1973, Sadat traveled to Riyadh to hold talks with Saudi King Faisal.

For Faisal, Saudi participation in an Arab–Israeli war through exercise of its oil leverage could bring his country many benefits. First, it could lead to a Saudi–Egyptian alliance, linking the richest and most powerful Arab states; conservative Riyadh, under attack from Arab extremists like Qadaffi, felt the need for Cairo's protection. Second, it would enhance Saudi prestige in the Arab world, lessening threats to the Faisal regime. Third, it might force the United States to reassess its Middle East policies, particularly its uncritical support of Israel. Fourth, the war would bolster oil prices, increasing Saudi revenues. Finally, should it be Allah's will, the war might end with the recuperation of Jerusalem.[33]

But Faisal was unwilling to cut his ties with the United States; in a hostile world, Washington was the most important guarantor of his security. Following his meeting with Sadat, he told both local oil executives and administration officials in Washington that he was under severe pressure to cut oil production. Many Arab oil producers had already halted production for one day on May 15, Israel's twenty-fifth birthday. Now there were proposals to reduce production on a steady basis until Israel withdrew from the occupied lands. He pleaded for a more even-handed Middle East policy in order to reduce the likelihood of war or unilateral producer cutbacks.[34]

But the opportunity for peacetime diplomacy had passed; war would now decide outstanding issues between the Arabs and Israel. On October 6, 1973, Egypt and Syria launched a coordinated attack on the Jewish state. The war caught Israel at a low level of preparedness, as reserves were at home for the high holiday of Yom Kippur. The first days of battle marked the greatest success ever achieved by the Arab states in their twenty-five-year struggle against the Zionists.[35]

The Arabs took the offensive on the oil front as well. On October 7, Iraq nationalized U.S. oil company interests in the Basrah Petroleum Company. At the same time, military operations halted shipment of over 1 million barrels of petroleum per day through various Middle East pipelines. Ironically, on October 8 OPEC began to hold talks with the major oil companies in order to renegotiate the Tehran agreement; these discussions "were rapidly overtaken by events. . . ." The following day, a Kuwaiti minister called for an emergency meeting of Arab oil producers.[36]

On the battlefield, Israel had managed to turn the tide against the aggressors within a matter of days. Afraid that its clients would once again suffer a devastating loss, the Soviet Union commenced massive arms shipments to Egypt and Syria during the second week of the war. Soviet naval battle groups in the Mediterranean were also strengthened. If Secretary of State Kissinger believed that détente would curb Moscow's fueling of regional violence, the Yom Kippur War taught him differently.[37]

Although the Israelis had gained the upper hand by mid-October, their human and matériel losses were enormous. Now that the Soviets appeared committed to a continuation of the conflict, the United States had no choice but to respond. On October 14, President Nixon ordered emergency arms shipments to Israel.

While the air bridge to Israel was being established, on October 16 the Arab oil ministers—under the banner of OAPEC (Organization of Arab Petroleum Exporting Countries)—met at the Kuwait Sheraton Hotel to determine their course of action. Their first move was to raise prices unilaterally, from $3.011 to $5.119 per barrel. As one Arab commentator remarked, "The situation presented an opportunity to make money and be a patriot at the same time."[38]

The following day, the gulf oilmen agreed on a 5 percent production cutback from September output levels. They further warned that similar cuts would occur monthly until the Israelis had withdrawn "from the whole Arab territories occupied in June 1967 and the legal rights of the Palestinian people restored." The Arab ministers said, however, that they did not wish to "harm any friendly state which assisted or will assist the Arabs actively and materially."[39]

The Europeans got the message loud and clear. Britain shipped arms to Arab states but not to Israel, with which it had contractual obligations. France supplied weapons to Libya and Saudi Arabia, which were then transferred to Egypt and Syria. The NATO countries, with the exception of Portugal, refused landing rights to American transport planes bound for Tel Aviv, but the Turks allowed the Soviet Union to violate their air space without a word of protest. The allies gambled that they could simultaneously win Arab favor, deny American military requests, and still maintain U.S. political support for NATO.[40]

Having successfully divided and conquered the Western alliance, the Arab oil ministers now allocated petroleum supplies on a political basis. On October 18, Abu Dhabi and Algeria announced a complete embargo of the United States; this action was followed by other oil producers between October 19 and October 21. From October 22 to November 2, the producers expanded the embargo to the Netherlands because of its alleged support for Israel. In contrast, France, Spain, and the United Kingdom were called "friendly" states and were allotted 100 percent of their normal supplies. The remaining "neutral" states (e.g., the Federal Republic of Germany) received cutbacks well above the announced 5 percent level.[41]

Even the threat of Soviet intervention on the side of the Arabs during the war failed to promote alliance cooperation; to the contrary, it produced further disunity. During the third week of October, as the Israelis were on the verge of destroying Egypt's Third Army, President Sadat proposed that a joint U.S.-USSR force enter the region to enforce a cease-fire. The United States opposed this idea but failed to win the support of its allies. According to one analyst, "Neither the British nor the French were willing to challenge Sadat and possibly risk economic reprisal."[42] In the absence of a positive U.S. response, the Soviets appeared intent on taking unilateral action, and intelligence reports stated that transport planes were ready to embark airborne troops.

On October 25, Secretary of Defense James Schlesinger ordered a worldwide alert of American forces, a decision reportedly made without alliance consultation. European governments were outraged by this unilateral act and suggested that they would prevent the United States from using NATO bases in the event of a superpower conflict in the Middle East. Ultimately the Soviets backed down, but as Kissinger told the allies on October 30, he "failed to understand" why they had acted "as if the alliance did not exist."[43]

In fact, the alliance response is not difficult to comprehend. The allies had two major objectives during the Yom Kippur War: first, to obtain oil supplies, and second, to avoid entrapment in a U.S.–Soviet conflict. This required that they curry favor with the Arabs while maintaining neutrality in the superpower dispute. European leaders did not act courageously during the war, but they acted in accordance with their immediate national interests. This is what statesmen are supposed to do.

Alliance Discord

The OECD oil committee remained on the sidelines during the Arab oil embargo. The committee's delegates recognized the "need for action," but mobilization of the oil allocation system required a unanimous declaration that an emergency existed. The members from Western Europe and Japan, however, feared taking steps that might further antagonize the Arab producers, and the only agreement reached was a pledge to provide monthly reports on oil stock levels. Debate on oil sharing degenerated as the members discussed various formulae. The French argued that any allocations should be based on "vital needs," while the United States backed allocations of "waterborne imports"; the Japanese suggested that emergency sharing be based on "total energy requirements." These alternatives were debated throughout October and November.[44]

With the failure of the OECD to implement emergency action, the response to the oil embargo became characterized by a scramble for supplies in which individual states sought special relationships with Arab oil producers. The Japanese courted Iraq, while the French proposed oil for arms deals with Saudi Arabia. British missions flew to Tehran, to be followed by the West Germans. The absence of a united consumer front allowed the producers to raise prices in dramatic fashion. According to a U.S. State Department study, spot market prices rose as high as $17 per barrel following the embargo. Official OPEC prices climbed from between $2.50 and $4.00 per barrel (depending on quality and location) during the third quarter of 1973 to nearly $12 by January 1, 1974.[45]

A superficial examination of energy markets during October and November 1973 makes the energy scramble hard to explain. The Arab production cuts were largely made up by increased output in Iran, Indonesia, Nigeria, Canada, and the Soviet Union, so that the total Western European shortfall amounted to only 4.4 million barrels per day. At the same time, stockpiles remained at the eighty day level; indeed, there is evidence that European governments were buying petroleum to add to stocks rather than for current consumption. Despite high stock levels and diverse suppliers—two important components of post-Suez energy policy—Western discord nonetheless won the day.[46]

A deeper analysis of consumer behavior helps to explain the outcome. Of greatest importance, the consumers were faced with a great deal of uncertainty regarding oil markets. It was unclear how long the embargo would last and how severe the shortages would become. Arab policies with regard to pricing and oil corporation activities were a mystery. Indeed, one OECD energy official has written that during the crisis "a reliable assessment of the world-market situation was not at all easy to obtain; information was . . . incomplete, contradictory, and confusing." Under such conditions, the allies were motivated to purchase petroleum, albeit at rising prices, for stocks. This was perfectly rational crisis behavior.[47]

The end of fighting on the Arab–Israeli front in late October failed to end the energy war raging among the allies. The French conception of a united Europe was shown to be vapid, as the European community turned its back on the embargoed Netherlands. In response, the Dutch threatened to halt natural gas exports to Paris. According to Robert Lieber, this threat promoted a "quiet resolution" of the Dutch situation. The European response to the oil crisis was based not on coordination with community or alliance partners but on an autarkic view of self-interest.[48]

Yet the interdependent nature of the oil industry made autarkic solutions difficult to realize. As the Europeans had learned during earlier crises, access to crude oil alone did not ensure energy security. The European continent was linked by a complex structure of pipelines, refineries, and distribution networks. France, for example, imported 7.2 million tons of some petroleum products in 1972 but exported 10 million tons of others; the United Kingdom imported 17 million and exported 15 million. Not a single European country enjoyed complete energy self-sufficiency.[49]

Accordingly, in the absence of specific international agreements to share available petroleum supplies, the oil industry assumed the allocation burden. This was not a happy task. As Shell and other firms had realized, a severe oil shortage would put them between severe producer and consumer country pressures from which they could hardly emerge unscathed. Allocation patterns that displeased producers could lead to oil company nationalization; ironically, the result on the consumer country end might be the same.

In the aftermath of the crisis, a U.S. government study of oil company allocations during the Arab embargo concluded it would

TABLE 7–2. World Oil Production During the Arab Oil Embargo (millions of barrels per day)

	1973				1974		
	Sept.	*Oct.*	*Nov.*	*Dec*	*Jan.*	*Feb.*	*Mar.*
Arab	20.8	19.8	15.8	16.1	17.6	17.9	18.5
Non-Arab	38.4	38.9	39.0	39.3	39.5	39.5	39.5
Total	59.2	58.7	54.8	55.4	57.2	57.4	58.0

Source: Federal Energy Administration, *U.S. Oil Companies and the Arab Oil Embargo* (Washington, D.C.: Government Printing Office, 1975).

have been "difficult to imagine that any allocation plan would have achieved a more equitable allocation of reduced supplies."[50] During the period of the embargo, which endured officially until March 17, 1974, U.S. imports of crude and product were cut by 12 percent, and Western Europe obtained 13.6 percent less. The Japanese managed to consume at preembargo levels for a variety of reasons, including increased supplies from East Asian producers, the quantities of oil en route to Japan at the time of the Yom Kippur War, and political pressure on the oil companies. Table 7–2 shows the sharp increase in non-Arab oil production to help meet the embargo shortfalls.

According to Robert Stobaugh, the oil companies allocated supplies on an equitable basis for three reasons: first, to lessen the risk of litigation and political reprisals by consumer governments; second, because of the desire to maintain an appearance of political neutrality; third, out of the necessity to maintain market flexibility, given the complex structure of oil production, refining, transportation, and distribution. In addition, the U.S. government apparently requested that the companies allocate supplies as fairly as possible. From a corporate perspective, all these reasons "are consistent with long-run profit-maximizing behavior."[51]

Unlike earlier energy crises, this time the United States failed to serve as alliance supplier of last resort. U.S. exports of petroleum products in 1973 totaled 229,000 barrels per day, only 7,000 barrels per day over the prior year's level; in 1974 they dropped to 218,000 barrels. The United States could no longer serve as market balancer. Indeed, imports grew to 35 percent of U.S. oil requirements in 1973; in 1967 they had met only 19 percent.[52] The hegemonic resource base of alliance energy security had eroded completely since the Six Day War.

But if the alliance leader perceived that energy security and alliance cohesion were inextricably linked, as it apparently had throughout the postwar period, why didn't it supply emergency supplies anyway, even if this meant domestic shortages? Isn't this the kind of action one would have expected from a leader? In searching for answers, we must examine both the domestic political and alliance dimensions of the energy shortage.

In terms of domestic politics, the Arab embargo was the first crisis to raise energy costs for American consumers and to confront them with oil shortages. One analyst of the embargo has suggested that the U.S. Congress would have been unwilling to share American supplies even if President Nixon had wanted to do so; the political ramifications were too great.[53] In short, even though certain domestic actors, such as Texas oil producers, would have benefited from emergency exports in a tight market, many others would have objected; crisis management had become costly to the United States.

In terms of the alliance, it is interesting to speculate on what would have happened had strategic interests converged rather than conflicted during the crisis. Had this occurred, President Nixon may have found a way to export more oil. But Western Europe and Japan refused to follow the U.S. leader; they had a different hierarchy of interests during the crisis (they cared more about oil than about Israel) and acted accordingly.

The Arab oil embargo precipitated the sharpest conflict in alliance relations since Suez. The United States had promoted the construction of an energy security structure after World War II, because it was aware of the strategic dangers posed by resource scrambles; in 1973 these dangers were fully revealed. As a leading European scholar observed, "It is doubtful whether the Alliance can again sustain extreme strains such as occurred during the October war. . . ."[54]

The European response to the Arab embargo placed the seed of doubt in the senior alliance partner, leading Americans to question the political value of the alliance. The only party to benefit from this schism was the Soviet Union, which learned that "by gaining control of the Persian Gulf it could gain control over Europe."[55] Without power or common purpose, the Western alliance hardly appeared as a durable guarantor of collective security.

The Energy Action Group

To Secretary of State Kissinger, there was no doubt that the United States must marshall what power it still possessed to reunite the allies. In a speech in London in December 1973, he proposed the creation of an "Energy Action Group of the industrial democracies—in effect, a consumer grouping to promote alternative energy sources and conservation and to negotiate with the producers." European reaction to the speech was, in Kissinger's words, "ambivalent." The allies continued to believe that energy security was best achieved by dissociation from the United States and the pursuit of bilateral deals. But the continuing scramble only exacerbated the most serious part of the energy crisis—the rapid increase in prices.[56]

Kissinger's boldness in attempting to promote alliance cohesion cannot be overstated. Not only did the allies have diverse strategies in the Middle East, but American leadership was being paralyzed by the Watergate and Spiro Agnew scandals. Agnew had resigned on October 10, 1973, and ten days later the world read of the "Saturday Night Massacre," in which Nixon fired several administration officials. By December, polls showed that 45 percent of the American people thought Richard Nixon should resign.[57]

In terms of international affairs, Kissinger's aborted "Year of Europe," announced in a speech in April 1973, had already made the allies wary of U.S. efforts to promote a new "Atlantic Charter" dominated by American interests. Indeed, the period prior to the embargo had been characterized by "extreme European–American antagonism." On economic issues, the Europeans remained distressed by U.S. international monetary and agricultural export policies. In terms of high policy, Walter Laquer notes that Washington had stopped consulting its allies on key strategic issues, including negotiations with the Soviet Union on conventional force reductions. The growing alliance divergences over

Middle East policy by early 1973 have already been mentioned; overall, the Arabs had demonstrated an excellent sense of timing if their objective was to exacerbate Western discord.[58]

Yet the United States remained the leading economic and military power within the alliance; Kissinger's negotiation of a Middle East cease-fire suggested that America's global role had not diminished. In contrast, "Europe was powerless to help shape the outcome of the Middle East problem." It lacked both the unity and capability to ensure the realization of its vital interests in the region and was left to making empty statements under the banner of the Euro–Arab dialogue.[59]

Economically, although the United States had increased its dependence on foreign oil, its indigenous resources remained great. Further, the United States was prepared to initiate an energy research-and-development effort that would dwarf any comparable program the European states—collectively or individually—could muster. Indeed, at the height of the Arab oil embargo President Nixon had launched "Project Independence," an energy program that he likened to the Manhattan Project and the Apollo space mission. As Kissinger stated in London, "The United States is prepared to make a very major financial and intellectual contribution to the objective of solving the energy problem on a common basis."[60]

As part of his lobbying effort to promote alliance cooperation, Kissinger made clear that the United States had far more options in the energy field than the other allies combined. Of interest, he also held out the promise of emergency exports of American oil supplies in the event of a future cutoff. As he wrote in his memoirs of Europe's choices:

> It was one thing to dissociate from us on Mideast diplomacy . . . it was another to advertise isolation on energy matters from a partner who, if it went the bilateral route, would have by far the strongest bargaining position and who would be needed as a supplier of last resort if there should be a major oil cutoff.[61]

On January 10, 1974, President Nixon formally invited the members of the OECD oil committee to attend a February meeting in Washington that would have as its objective the establishment of a consumer country energy task force. The immediate response was not altogether positive. The European community debated whether or not to accept, as France initially argued against the idea. The Euro–Arab dialogue, the French asserted, and bilateral efforts to improve relationships with the producers offered greater promise for energy security than consumer cooperation with the pro-Israeli United States. But this line was not altogether convincing to the other Europeans. The energy crisis could not be solved in isolation from broader security, trade, and financial issues. Even if the community could somehow obtain preferred access to petroleum supplies, it was unlikely that the member states could collectively overcome these larger problems in which the oil crisis was embedded.[62]

The allies were ultimately driven to accept the Kissinger–Nixon energy initiative owing to the failure of their bilateral efforts to ensure *long-term* oil security. The official prices posted by OPEC had quadrupled for all buyers, and spot prices were even higher. The energy crisis was beginning to look like a chronic problem, not just a temporary shortfall. The perception that the world had entered an era of high-priced oil, and consequently that long-term solutions were needed, prompted the allies to accept Kissinger's invitation to an energy conference. Somewhat begrudgingly, the NATO allies and Japan assembled in Washington on February 11, 1974.[63]

Kissinger described the Washington energy conference in his memoirs as "a strange event. It was a meeting of allies, but it had something of the character of a clash of adversaries. . . ."[64] The secretary's relationship with French foreign minister Michel Jobert was particularly strained. But this should not have been a surprise. For the French, the Arab oil embargo had presented a crisis of opportunity. It permitted France to exploit its self-image of being not only the European community's leader but also the developing world's best friend and an independent force in Middle East diplomacy. France was backing this strategy by committing additional armed forces to the region, assigning a carrier battle group to the Indian Ocean. France also hoped to resolve its historical grudge against Britain and the United States, the two states that it believed had conspired to keep French oil interests out of the Middle East after World War II. As Louis Turner has written, "The whole history of France's involvement in oil has been a story of bitter struggle against Anglo–Saxon dominance." Its pro-Arab policies were a "calculated attempt to regain the influence and oil which France had lost in the . . . 1940s." Simply stated, France had its own agenda in the Middle East.[65]

Franco–American discord during the energy conference placed the European community ministers in a bind, since several of them were sympathetic to Jobert's stance. Britain was notably supportive of an independent European approach to energy and Middle East problems. The West Germans, in contrast, who were represented at the conference by Finance Minister Helmut Schmidt, were more responsive to the American plan to establish an energy coordinating group. The differing views of the European community members and Japan threatened to paralyze the conference. The deadlock was broken, in part, by American linkage of energy to other issue areas.

Kissinger himself had been careful not to be heavy-handed in linking energy to alliance military security. But Richard Nixon was less tactful. At a dinner for the delegates, Nixon said that "security and economic considerations are inevitably linked and energy cannot be separated from either." He noted growing protectionist and isolationist pressures in the Congress, pressures he vowed to fight so long as there was "cooperation in security . . . cooperation in trade . . . cooperation in developing our sources of energy and in acquiring the energy we need."[66]

There is no doubt that Nixon's tacit threat to remove troops from NATO was taken seriously by those assembled. Alliance relations had sunk to a low point during the Arab embargo, as the Europeans refused to cooperate with Washington's resupply of Israel, or Kissinger's efforts to prevent a Soviet-sponsored cease-fire. In Walter Laquer's words, the Europeans had "behaved as if they wanted to provide support to the American critics of U.S. military involvement in Europe, such as Senator [Mike] Mansfield."[67] Given the weakened Nixon presidency, it was quite conceivable to the European delegates that anti-NATO sentiment in Congress could win the removal of some American troops.

Ironically, Nixon's linkage of energy and security pointed to the weakness of the alliance leader. In earlier times, the United States had possessed the fuel resources needed to win alliance cohesion during energy crises, as it acted as supplier of last resort. Now, Washington had to use negative sanctions in the military security realm—specifically, the threat of a partial withdrawal from

NATO—to promote a coordinated energy policy. In the short run, this linkage would help produce desired results; with the exception of France, the Europeans agreed to establish an energy coordinating group. But it was unclear over the longer term whether the use of threats would promote alliance solidarity or undermine it further.

The energy–security link may have been necessary to bring about an agreement in Washington, but it was not necessarily sufficient. The United States still possessed vital resources in the energy field that other countries wanted. The Germans and British were eager to cooperate in energy research and development, and the Italians sought U.S. help in establishing an oil market information system and broad surveillance of energy market activities. Although Washington was no longer a reliable oil supplier of last resort, it could still provide its allies with tangible benefits.[68]

The Energy Coordination Group held its first meetings at the ministerial level in March, and working groups were established in such areas as oil sharing, research and development, conservation, and government relations with oil companies. The group continued its work over the summer, and in September the International Energy Program (IEP) was signed in Brussels. It called for the creation of an International Energy Agency (IEA) as part of the OECD. The IEA began operations in November 1974, and it is discussed in the following chapter. With the IEA's establishment, the OECD oil committee, which had helped to manage so many postwar oil crises, ceased to be an important part of the alliance energy security structure.

Conclusion

The world oil market was completely transformed during the years following the Six Day War. Changes occurred on both the demand and supply sides of the equation, as producer country militance was supported by U.S. profligacy. Indeed, market changes cannot be viewed in isolation from the energy policies that gave them rise. Oil price controls in the United States and the end of the mandatory oil import quota were two notable decisions that favored foreign oil at the expense of high-cost domestic production.

The collapse of U.S. spare capacity, along with environmental policies that stymied the development of the Alaskan oil fields and the utilization of alternative fuels, created an energy crisis in America several months before the Arab oil embargo. The Western allies were well aware of the new market conditions, and they attempted to build stocks and diversify supplies in the event of a future crisis. These national policies failed to provide energy security during the embargo, because, ironically, they demanded a coordinated consumer response to succeed.

The changes on the energy front were more than matched by alliance conflicts in the Middle East ambit. Washington's uncritical support for Israel posed a problem for oil-dependent Europe and Japan, and the allies accordingly made strenuous efforts to distance themselves from U.S. policy. While some of them may have been rewarded during the embargo by getting on the Arabs' "friendly" list, they quickly learned that it was impossible to become isolated from the global price shock. At the same time, superpower conflict in the Middle East could spread to the central front, entrapping the allies in a conflict which was not of their making.

The end product of these events on the energy and security fronts was alliance discord. The absence of emergency supplies from the United States, coupled with alliance divergence over Middle East policies, created a conflictual situation which doomed Western cohesion. Alliance policy coordination could not occur in the absence of hegemonic power and common strategic purpose.

After the immediate oil shock ended, however, the United States proved capable of reasserting its leadership. Over the long term, the Europeans and Japanese recognized the futility of going it alone. Energy problems were too bound up with issues of trade, finance, and international security. Autarkic solutions could not be pursued by small states at an acceptable price.

Thus, the oil embargo of 1973–1974 was characterized by both continuity and change. In the end, the International Energy Agency was established, with responsibilities similar to those of the OECD oil committee. But in the future, alliance policy coordination during energy crises must occur in the absence of a supplier of last resort. The ability of the allies to adapt will be examined in the following chapter.

8

The International Energy Agency: 1974–1980

> No area of [energy] policy has been more disappointing than consumer cooperation.
>
> HENRY KISSINGER[1]

The International Energy Agency (IEA) was created in 1974 in direct response to the Arab oil embargo. Why was a new alliance organization deemed necessary? For over twenty years, the energy and oil committees of the OEEC and OECD had competently dealt with alliance energy matters. They had been effective in coordinating a multilateral response to the oil crises of 1951, 1956, and, to a lesser extent, 1967. It could hardly have been argued in 1974 that alliance cooperation was stymied by the lack of an appropriate organization.

This chapter treats the IEA during its first decade. The alliance has faced in this period the Iranian revolution of 1978–1979, the Soviet invasion of Afghanistan, and the enduring Iran–Iraq war launched in September 1980. These cases provide the empirical background against which collective action in the energy area can be assessed.

I will argue that the establishment of the IEA signified a fundamental change in alliance energy crisis management. Prior to 1973, the United States had unilaterally balanced energy markets at times of shortage. By maintaining domestic spare production capacity with a protected market and high prices, it bore most of the costs associated with alliance energy security. But as the costs associated with this policy became unacceptably high, new strategies to share the burden of energy security had to be developed. The IEA has sought to manage the collective action problems that have emerged in the energy area with the decline of U.S. hegemonic power.

The IEA: Structure and Functions

The International Energy Program (IEP) was signed by sixteen countries in September 1974. It called for the creation of an International Energy Agency as part of the OECD. The IEA began operations in the Paris headquarters of the OECD in November 1974.

The only members of the OECD that have rejected IEA participation are Finland, France, and Iceland. The French attitude toward the IEA was discussed in the previous chapter; those of Finland and Iceland also deserve mention. Prominently, both of these states were heavily dependent on the Soviet Union for their oil supplies. Iceland has traditionally received more than 50 percent of its petroleum from Russia, largely on the basis of fish-for-oil barter deals. Finland, wedged against its superpower neighbor, is the single largest OECD importer of Soviet oil. The Kremlin has embargoed trade to both of these countries in the past to express dissatisfaction over their foreign policies, as in 1949 when it broke trade relations with Iceland while Reykjavik pondered NATO membership. Further, in terms of narrow economic interests, the Soviets have provided Finland and Iceland with oil on attractive terms, making IEA membership even less attractive.[2]

The IEA is run by a secretariat consisting of the executive director and assisting staff members, and specific programs are developed by the four standing groups which focus on (1) emergency questions, (2) long-term cooperation, (3) the oil market, (4) relations with producer and other consumer countries. There is also a committee on energy research and development, with the status of a standing group. The IEA's work is overseen by a governing board composed of energy ministers of member states, and during an oil crisis a management committee of senior energy officials is formed. The major oil companies participate in the IEA through the International Advisory Board (IAB), though this group's involvement has been somewhat constrained by U.S. antitrust laws, which preclude American firms from jointly discussing certain issues (e.g., oil prices) in the absence of an emergency declaration by the President. Any "decision" concerning IEA activities requires a 60 percent majority vote of the participating countries; votes are weighted on the basis of national oil consumption. The United States, for example, possesses fifty voting weights, or nearly one third of the IEA total, and this gives it substantial power in all deliberations.[3]

The IEA's foremost task in its early years was the establishment of an energy emergency program, which amended procedures earlier established by the OEEC and OECD oil committees. In order to activate the program, the secretariat must first make a "finding" that an oil shortfall of 7 percent or more exists within a member state or among a group of states; any participating country can also request the secretariat to consider making a finding. This 7 percent figure is an artifact; it represents the overall OECD oil shortfall that existed during the 1973–1974 Arab embargo and had no deeper significance. A finding is made only after consultation with the IAB, to allow the companies an opportunity to respond on an ad hoc basis if this is deemed appropriate. If the secretariat makes a finding, it reports this to the IEA's management committee, which confirms or denies the decision on the basis of a simple majority vote. Even a confirmed finding, however, can be overruled by a 60 percent majority vote of IEA members. Should the finding be confirmed, the emergency program is activated. This process, from the initial finding of a shortfall to implementation, can take nearly a month.[4]

National energy policies make up the first component of the emergency management system. Participating countries must restrict oil demand and begin to draw down stocks from strategic petroleum reserves (SPRs); under the IEP states must hold—in public or private sector hands—reserves equal to 90 days' consumption. Only after reserve drawdowns have reached 50 percent of the members' total holdings is a second finding made for allocation of crude oil supplies. This leads to another set of potentially long and complicated negotiations among states and companies. If this second finding is affirmed, the oil companies are ordered to allocate available petroleum supplies on the basis of a formula that gives each country, including the United States, a "supply right"—a right to a specified percentage of available oil.[5]

The IEA has never made a finding since its inception, even though a couple of countries have requested the secretariat to launch finding investigations, as will be discussed in a later section. A parsimonious explanation for the absence of emergency action is that the consuming nations have not faced an oil shortage as bad as the one of 1973–74; the IEA scheme has a specific 7 percent trigger. Some observers, however, see the agency's failure to declare a finding as an admission that its emergency program is ineffective, that it cannot be counted upon to coordinate alliance crisis behavior.[6] Skeptics have asserted that in an emergency the IEA's program simply "would not work."[7]

I would argue that the IEA, recognizing the difficulty of crisis management in the absence of a hegemonic leader, has in fact focused its efforts on *national* energy security measures. In this regard it should be recalled that the first tier of the IEA program consists of demand restraint and stockpile withdrawals. The latter has long been viewed as critical to energy security; as M. A. Adelman wrote in 1967, "the only way to cheap and secure fuel is to stockpile oil. . . ."[8] By placing oil stockpiles at the heart of the emergency program, the IEA attempted to create an effective deterrent in order to prevent politically induced crises in the future. As I will show in the following sections, this approach has proven inadequate to guarantee alliance energy security.

The second major focus of IEA activity has been the promotion of multilateral energy research and development projects; American promises to lead in this area were one reason why the allies ultimately agreed to join the agency. By the late 1970s, the IEA had working parties involved in the study of energy conservation, coal technology, nuclear safety, and alternative energy development. To date, the lack of progress in this area has testified to the difficulties inherent in multilateral cooperation in research and development. The impediments have included corporate reluctance to share proprietary information, conflicting national competitive strategies, and differences in business-government relations within the IEA member states. On a theoretical level, the problem is of the collective-goods variety; states are unwilling to fund the public good of knowledge, since it allows free riders to obtain disproportionate benefits.[9]

The third, and perhaps most important, objective of the IEA has been to establish an energy market information system. In the words of a former IEA employee, the agency's "information systems on energy markets are without equal."[10] The information system has greatly increased the transparency of energy markets, making it easier for states to follow energy flows across borders. This information is vital to the agency's confidence-building measures, and its provision may help to explain why the IEA has endured despite the lack of decisive action during the energy crises it has faced.

The IEA has used its information system in yet another way to promote policy coordination. As a price of membership in the organization, each state has agreed to allow the secretariat to conduct an annual, energy policy audit. These published surveys describe the precise measures that member states have taken to bolster their energy security. "Peer pressure" has been placed on IEA states to adopt similar policies, and the energy policy process within nations has become more transparent.

In comparison with the OEEC and OECD oil committees, the IEA exemplifies both continuity and change. The IEA's guiding principles are to develop "secure oil supplies on reasonable and equitable terms" and, in the event of an oil shortage, to share available supplies on an equitable basis. These principles resonate with the past. The rules for participation, however, are much more complicated, and include not only stockpiling and the promotion of oil diversification strategies, but participation in energy research and development projects and support for energy conservation programs. The IEA's voting system is also more complex than the one that informed its predecessors; consensus has given way to weighted voting.

Although the IEA is a more ambitious undertaking than the earlier oil committees, this does not necessarily mean that alliance energy security has been strengthened by its creation. Since the early 1970s, the United States has lacked the energy power that once fueled postwar alliance security, and it has become a substantial oil importer. Further, the United States has continued to support Israel in a way that the allies perceive as detrimental to their oil security. The IEA could not ameliorate the tensions in alliance relations caused by these facts of international life.

Nonetheless, the establishment of the IEA stands as a testimony to Henry Kissinger's statesmanship at a difficult moment in alliance history. While Kissinger has expressed disappointment over the IEA's work, the agency's ongoing maintenance problems should not overshadow the effort that went into its creation. In linking energy and alliance security, Kissinger and President Nixon forced the allies to reflect on their ability to shape great events without a hand from Washington.

Oil Market Changes After the Embargo

The International Energy Agency was established within the context of the Arab oil embargo of 1973–1974 and the oil market structure of the early 1970s. In the aftermath of the embargo, the process of market transformation that had been unleashed by Qadaffi was accelerated. OPEC members increased their control over the market by obtaining majority interests in producing ventures, and in turn they gained control over production volumes and, to a lesser degree, control over destinations through the use of export restrictions. The spot market increased in volume, permitting buyers and sellers to trade oil in the absence of long-term contracts with one or more of the seven sisters or the producers. The end of long-term agreements increased oil price volatility, as prices responded quickly to any number of political and economic variables. The security implications of these changes were ambiguous; the loss of corporate power was certainly a source of anxiety to Western policymakers, but as the oil producers moved downstream into refining and marketing it was believed they would become more moderate in their dealings with consumer governments.[11]

OPEC members themselves were divided over how best to respond to the economic gains produced by the embargo. The moderate Saudis were concerned lest further price hikes undermine the international economy. The price hawks, led by Algeria, were far less interested in maintaining the capitalist status quo. As journalist Ian Seymour has written, all OPEC members knew that some price erosion must occur after the price shock of 1973–1974; "The question was: how far, how fast, and where to draw the line?" The ensuing years saw near continuous debate on this topic.[12]

OPEC's problems were complicated by the decline in petroleum demand that occurred after the embargo. The price shock had catalyzed a host of economic problems in both the developed and developing worlds, and it unleashed that terrifying combination of inflation and stagnation—"stagflation." At the same time, the industrial countries began to turn away from OPEC sources, as oil from the North Sea and Alaska's North Slope came on stream. Crude oil production among OPEC producers fell from 31 million barrels per day in 1973 to 27 million in 1975.[13]

The market changes of the 1970s undermined but certainly did not annihilate the position of the multinational corporations in the oil industry. Although OPEC producers claimed majority interests in most of the producing companies, the firms continued to hold the operating contracts, and they supplied most of the personnel and infrastructure support for oil operations. And while government-to-government and other direct deals claimed, in 1976, 20 percent of OPEC production, this left 80 percent still in the hands of the multinational distributors. Further, the majors had displayed their flexibilty by seeking new sources of oil outside OPEC and by investing in the development of alternative energy sources.[14]

Ironically, the position of the oil firms *within* the consuming countries was no easier. In response to the public outcry over abnormal corporate profits following the Arab oil embargo, investigations of oil company activities were launched in Western Europe and the United States; the hearings held by the U.S. Congress, under Senator Frank Church, now form an important part of the literature on international oil. The governments of Germany, France, and Japan created new national oil companies to seek fresh supplies.[15] The multinationals thus faced the difficult task of satisfying their OPEC and OECD critics.

Perhaps the most significant post-embargo factor in the oil markets, from the alliance perspective, was the continuous growth in U.S. oil demand and imports. U.S. oil consumption rose from 783 million tons in 1974 to 889 million in 1978, and of this latter amount 44 percent was imported—7 percent above the 1974 figure. Even more disturbing, 35 percent of U.S. imports in 1978 derived from Persian Gulf producers, up from 29 percent in 1973.[16] Not surprisingly, leaders of the industrial democracies expressed sharp criticism over U.S. energy policy, and at the Bonn summit of 1978 they demanded import restraint as the price Washington must pay for international economic cooperation.[17]

Despite high stock levels, soft prices, and the existence of the IEA, uncertainty about the future characterized world energy markets in the late 1970s. There was a pervasive belief, fueled largely by a CIA study released in 1977, that the world faced a chronic energy shortage that must lead to crisis. The agency predicted that demand for OPEC oil would outpace the growth in supply, leading to shortages and higher prices. To complicate the international situation, the CIA argued that Soviet oil production was peaking and that in the 1980s Moscow and its clients would be forced to purchase petroleum from OPEC producers. Academic analysts echoed the CIA report; in a widely cited piece in *Foreign Affairs,* Dankwart Rustow argued that physical shortages of oil would soon plague the alliance, causing a price jump comparable to that of 1973–1974.[18]

But those who foresaw energy shortages were worried about the 1980s, not the late 1970s. The threat of another Arab–Israeli war had receded into the background, following President Sadat's dramatic visit to Jerusalem in November 1977 and the subsequent bargaining which led to the Camp David accords of September 1978. The producers seemed less homogeneous, as Saudi Arabia and the United States had developed a "special relationship" while Libya and Algeria appeared increasingly isolated. At the same time, the consumers were building stockpiles and developing alternative energy sources and suppliers. The late 1970s were apparently providing the allies with sufficient time to prepare for the next crisis.

The Iranian Revolution

In mid-1978, oil market disturbances occurred in an unlikely place—Iran. The contract between Iran's oil workers and the country's foreign oil consortium expired in July 1978, and wide differences between the parties quickly emerged. As negotiations stalled throughout the summer, more militant action was employed by the workers. Strikes began in the Ahwaz oil fields in September, later spreading to Abadan; on October 13 the great refinery was shutdown.[19]

By some accounts, the oil workers' strike catalyzed a loose coalition of anti-Shah activists to hit the streets. Ironically, these activists took comfort from the Carter administration's human rights policies, believing that Washington would hear their cries. This proved to be a tragic mistake. On September 8, 1978, the Shah's imperial guard fired on a mass demonstration, killing as many as 4,500 people. This day of infamy would live as "Black Friday" in the hearts and minds of anti-Shah forces.[20]

The extent to which the Iranian revolution of 1978–1979 was "organized" by internal or external elements remains a point of dispute among academic analysts. Indeed, one of its telling characteristics was an apparent spontaneity, as demonstrators charged the streets not only on Islamic holy days but also in support of the oil workers and those whose relatives were missing in the Shah's prisons. By January 1979 it was clear to U.S. and Iranian officials that the Shah no longer served as a legitimate leader. In February, the Ayatollah Khomeini returned to Tehran from his exile in Paris to establish a new Islamic republic.

The oil field disruptions of autumn 1978 were quickly translated into market hysteria. Spot market prices soared to 20 percent above official OPEC prices, surging past $14 per barrel. "It's pure panic," one oil trader told the *Petroleum Intelligence Weekly* in mid-November.[21]

Buyers were entering the market for several reasons. First, arbitrageurs saw the opportunity for quick profits by buying and selling in the spot market. Second, refiners who depended on Iranian crude scrambled to find alternative suppliers. Finally, governments—notably the United States—remained in the market in order to add to stockpiles. Taking advantage of the panic,

OPEC raised its official prices by 14.5 percent during Christmas week 1978, but this hike paled in comparison with the unofficial increases found in the spot market.[22]

The International Energy Agency observed the initial oil scramble from the sidelines. In mid-November, IEA Director Ulf Lantzke assured member states that the Iranian shortfalls would be made up by increased production elsewhere. Stocks were at high levels, so there was "no serious reason for concern." In any event, the loss of oil supplies was well below the 7 percent threshold that triggered the emergency system.[23]

But individual member states were nervous. Iran had supplied IEA members with over 16 percent of their crude imports, or over 3 million barrels per day. In the short run, this amount would have to be made up by other OPEC producers, notably Saudi Arabia. It was unclear whether the Saudis would provide the needed amounts and, if so, how long it would take to readjust production and transportation schedules. Once again, the consumers faced a shortage the duration and scope of which could not be predicted. The Iranian revolution had a shapelessness and mysteriousness about it, which compounded the difficulty for those analysts who were attempting to discern if and when Iranian oil exports would regain their historic levels.

Faced with uncertainty, the immediate response of the Western allies to the Iranian shutdown was one of panic and a scramble for available supplies; the IEA was totally ineffective in calming consumer fears, and the United States proved powerless to overcome the situation. Richard N. Cooper, who served as Undersecretary of State at the time, recalled that states were willing to "pay any price" for oil supplies.[24]

The revolution demonstrated that Washington was now impotent when it came to crisis management. Unlike 1953, when the CIA overturned a radical government in Iran, in 1979 it stood by as a group of "students" took its citizens hostage. In terms of energy power, the absence of spare production capacity meant that it could do nothing for its allies. Consumer cooperation was an empty shell.

In the absence of IEA action in late 1978 and early 1979, the oil companies themselves unilaterally allocated oil, as they had done during the Arab embargo. Exxon, for example, attempted to im-

TABLE 8–1. Oil Company Supply Losses, First Quarter, 1979

Company	*Percent of Total Supplies*	*Volume (Barrels per Day)*
Exxon	10	400,000
British Petroleum	45	1,400,000
Compagnie Française des Petroles	15	220,000

Source: *Petroleum Intelligence Weekly*, February 12, 1979.

plement the IEA formula as closely as possible, sharing oil on the basis of available supplies including domestic consumption levels. As Table 8–1 illustrates, the company faced a worldwide 10 percent shortfall in its normal supplies (approximately 400,000 barrels per day), and it allocated this cut evenly across its markets, hoping to avoid the charge that it treated particular consumers unfairly. Other companies also adopted the IEA allocation scheme as the crisis wore on; in the words on one oil executive, "An IEA type allocation approach could be a marvelous defense when politicians come gunning for so-called oil industry profiteers."[25]

Yet by the early spring of 1979, the corporate approach to fuel rationing was inadequate in the eyes of certain IEA member states. For reasons of domestic politics, some of these states had maintained domestic price controls and were now confronting shortages approaching the IEA's 7 percent trigger, as marginal oil supplies flowed to freer markets where higher prices were found. Recognizing that the problems would not go away, the IEA secretariat had no choice but to get directly involved in crisis management.

At a meeting on March 2, the members of the IEA's governing board declared their readiness to reduce oil demand by 2 million barrels per day, or 5 percent of overall oil consumption, the amount that had been lost from the market as a result of the Iranian uprising. The member states agreed to encourage fuel switching, conservation, and increased domestic energy production and to enact demand restraint policies. Unfortunately, these policies were undercut by the absence of a precise timetable. Further, the program developed at the ministerial meeting was entirely voluntary.[26]

Many states subsequently failed to comply with the voluntary agreements they had made. Analytically, the problem was of the "free-rider" variety. In the case at hand, if the United States, for example, had limited its oil imports by 500,000 barrels per day, this would have placed more oil on world markets and, in turn, dampened prices. Since the United States would have taken this action unilaterally, other states could do nothing and still enjoy the lower prices that resulted from American policy. The other consumers would become free riders on the collective good of lower oil prices. As free riders, they would have no incentive to "contribute" to cooperative policy measures.

But throughout the postwar period the United States had permitted its allies to act as free riders on a variety of political, military, and economic issues. Why did it refuse to do so during the Iranian crisis? To put the question another way, why did the United States refuse to lead?

Indeed, the United States not only refused to lead, but it added to the oil problem by subsidizing imports of heating oil and diesel fuel from Caribbean refineries, placing additional pressure on product prices. According to Richard Bissell, "To the Europeans, this action epitomized American unilateralism . . . the subsidies provided a focus for the European Community's previously undefined outrage (over U.S. energy policy) . . . ."[27] During the first quarter of 1979, U.S. oil demand actually rose by 1.4 percent over the previous quarter, to a record 20.33 million barrels per day.[28]

The American response to the Iranian crisis cannot be understood without reference to domestic political factors. In early 1979, President Carter attempted to fashion a leadership role, urging domestic refiners to refrain from purchasing spot market oil and warning that he would punish them by regulating gasoline prices, thus making it impossible for them to recoup their higher costs. The president outlined an ambitious energy conservation program, and on March 1 he sent a standby gasoline rationing plan to Congress. Through conservation and fuel switching, the president sought to lower U.S. demand for oil and contribute to price stabilization.[29]

The Carter program was severely criticized across the country. On May 16, 1979, California governor Jerry Brown flew to the White House to complain of gasoline shortages in his state which were creating consumer panic. According to economist Richard

FIGURE 8–1. Copyright 1979 by Herblock in The Washington Post.

Mancke, "Almost simultaneously, the Department of Energy reversed its previous policy and began to urge domestic refiners to purchase more oil on the spot market." The president had been stymied by effective domestic opposition to his policies. Congress approved only one of his four energy-saving requests, and these curbs on heating and air conditioning use would only result in oil savings of 200,000 barrels per day.[30] Herblock captured U.S. policy in a cartoon drawn in the early spring of 1979 (see Figure 8-1).

The IEA's governing board meeting of May 21–22 was characterized by a high degree of alliance mistrust. It was clear that several IEA members were failing to meet their conservation targets and that spot market activity was driving oil prices way above official OPEC levels. Despite a common interest in containing the economic shocks caused by the Iranian revolution, the allies were defecting from their agreements and making a bad situation much worse.

To complicate matters, at the May meeting Sweden and Denmark formally requested activation of the emergency oil-sharing system. After a cursory investigation, the secretariat refused to act, and instead it asked the oil companies to make some additional oil available to these markets. One student of IEA activities during this period states that the IEA feared that "activation of the allocation mechanism might cause panic hoarding." This, of course, was at odds with the very purpose of the emergency system, which was to prevent a scramble for oil.[31]

The IEA's failure to activate the emergency allocation system raised further questions about the viability of crisis management in a nonhegemonic alliance. The problem was made all the more complex because IEA member states had very different domestic energy policies. Sweden, for example, had maintained domestic price controls throughout the Iranian crisis, and some observers felt this was why it suffered abnormal supply shortfalls during the spring of 1979. So long as states took adequate crisis prevention measures, they were free to adopt any domestic policies they wished. But they could not expect to receive marginal supplies during a shortage if their price structure was uncompetitive.[32]

The aftermath of the Iranian revolution and a new OPEC price hike provided the setting for the Tokyo summit of June 28–29, 1979. The meeting opened with French president Giscard D'Estaing delivering some sharp criticism of U.S. energy policy. D'Estaing pointed to Washington's failure to save even a drop of fuel, and noted rising U.S. oil imports, which by June 1979 were almost 27 percent above 1973 levels. The Europeans sought an American commitment to freeze imports at current levels through 1985.[33]

After two days of bickering, a compromise was reached on import ceilings. The United States, Canada, and Japan set specific import targets out to 1985, while the Europeans established an overall EC target level. Despite all efforts to convince the press that a considerable degree of consumer country unanimity had been achieved, the Tokyo agreements were met with considerable skepticism. Analysts had seen the allies in action in 1973 and 1978–1979, and there was still no indication that cooperation in the midst of crisis could be achieved. The summer of 1979 provided the allies with a respite to their energy problems, but not a solution.

Summary

Two economists have neatly summed up the alliance experience during the Iranian revolution. "In 1979," they wrote, "the industrialised countries of the OECD inflicted on themselves one of the most disastrous events in their economic history."[34] A small supply shortfall caused a 150 percent increase in oil prices, with severe economic implications for the countries of North and South alike. For much of the crisis, the IEA sat on the sidelines, and when it entered, the best it could do was win a voluntary agreement on energy savings that few of the member states actually adopted. The United States failed to provide even a modicum of leadership, as the president bowed to domestic interests and allowed the country to scramble for supplies along with all other consumers.

The crisis suggests that the IEA was unprepared to manage an oil shortfall, no matter how slight. It had been established to deter, but not fight, the last war. When national measures yielded to alliance crisis management, the organization proved inept. In the words of Robert Weiner:

> the Iranian experience demonstrated that the IEA was not well equipped to deal with the problem of oil market "tightness"—in this case, supply reduction below the emergency threshold level. IEA countries continued to build their stockpiles during the period of Iranian shutdown by buying in the spot market. . . . Spot market prices soared, and posted prices followed shortly thereafter. Panic was the order of the day.[35]

It would be unfair, however, to pin the IEA's shortcomings on the organization itself. After all, implementation of IEA rules cannot take place without support from the member states. A coordinated approach to the shortage, through stockpile releases, demand restraint, and ad hoc allocations, could have prevented panic and ameliorated the price shock. But in the absence of leadership it proved impossible to establish common policies.

More than any other postwar crisis, the Iranian revolution points to the critical role hegemonic power has played in maintaining an *effective* alliance response to energy shortages. During the Arab oil embargo of 1973, the allies were divided not simply because the United States failed to ship oil, but also because of strategic differences over how the Arab–Israeli conflict should be managed. In addition, the allies had differing perceptions regarding the Soviet threat to the Middle East. In 1979, in contrast, there were no fundamental strategic conflicts among the allies, even though the United States was held largely responsible for the Shah's demise.[36] The IEA *endured* the crisis because the allies continued to have convergent interests in the energy area, but it was *ineffective* because power and leadership were absent.

It cannot be said, however, that President Carter had overlooked the alliance dimension of the crisis. To the contrary, alliance issues were very much on Carter's mind.[37] But the cost of alliance leadership was unacceptably high to domestic actors, and presidential initiatives were blocked as a result. U.S. domestic politics played a decisive role in energy policy formulation during the Iranian revolution, constraining the president's freedom of action.

Yet the negative outcome of the Iranian crisis did not mean that consumer cooperation was a hopeless endeavor. The allies recognized that if security was to be achieved in the context of unstable oil markets, it must be the product of coordinated policies. The question was whether the allies could learn from the experience of 1978–1979 and then apply those lessons to their common energy problems.

Afghanistan and Alliance Energy Security

The Iranian revolution presented the alliance with the specter of Islamic fundamentalism, a world view that was implacably hostile not only to Western society but also to many of the conservative regimes in the Persian Gulf. During the late 1970s it thus appeared that infectious ideological forces had supplanted the Arab–Israeli conflict as the primary threat to energy market stability. While every alliance member shared an interest in containing the spread of fundamentalism, it was unclear whether they should associate with the United States in that endeavor; Washington, after all, had been linked to the Shah. At the same time, the United States did not appear to have the politico-military or economic capability to stop Khomeini.

Even as alliance economies were still feeling the aftershocks of the Iranian revolution, yet another threat to energy security emerged on the horizon. In late 1979, Western intelligence sources indicated that the Soviet Union would send military forces into Afghanistan in support of the Communist regime which had been in place since April 1978. Resistance to the regime had mounted, particularly in the rugged rural areas outside Kabul, and the government appeared unable to quell local disturbances. On December 27, 1979, the Soviet invasion of Afghanistan began.[38]

To the United States, Moscow's drive southward placed the problem of energy security within the East–West context. A sideline observer of the invasion might have been forgiven if he or she expected to see a strong, collective alliance response, since a common fear of Soviet aggression was what bound the allies together in the first place. A Soviet threat to the Persian Gulf confronted the allies with the gravest "out-of-area" conflict imaginable.

In the early days of 1980, Deputy Secretary of State Warren Christopher flew to London to meet with several NATO ambassadors. The postwar doctrine of containment had apparently failed to prevent Soviet moves toward the Persian Gulf. A new expression of collective purpose was clearly needed.[39]

But Christopher's talks were soon derailed, as the allies disputed among themselves and proved unable to find common ground. The intra-alliance conflict exemplified NATO's out-of-area problem. While NATO members were bound by fear of the Soviet Union, collective action was limited to specific geographic coordinates defined in the North Atlantic Treaty. Alliance action outside this area had long been the source of western discord, as at Suez in 1956. The allies recognized that they had numerous unilateral interests outside the NATO zone and that these interests might not be shared by other member states. Accordingly, throughout the postwar period they tried to finesse the out-of-area problem, dealing with it on an ad hoc basis.

In a dramatic attempt to reassert U.S. leadership, President Carter announced on January 23, 1980, a policy that would soon be labeled the Carter Doctrine. "Any attempt by outside force to gain control of the Persian Gulf region," the President declared, "will be regarded as an assault on the vital interests of the United States of America, and such an assault will be repelled by any means necessary, including military force." Energy security was now a military security problem.[40]

Carter's response to the Afghanistan crisis effectively entrapped the allies. At the time the president announced his doctrine, the United States lacked the military forces required to make good his commitment to Persian Gulf defense. The Administration's call for a rapid deployment force meant that troops and ships must be withdrawn from NATO and other foreign duties. In addition, European bases would be needed as staging areas for Middle East operations. One way or another, the allies had been drawn into the superpower rivalry now raging in southwest Asia.[41]

The allies were entrapped by the Carter Doctrine in yet another way. Although the president had announced his speech as a unilateral initiative, it was clear to all observers, in the United States and abroad, that the Congress would not send American boys to die for Middle East oil, the bulk of which was exported to Europe and Japan. If the allies were concerned with their energy security, they would have to support the U.S. military effort, either by making up for the American troops shifted out of NATO, by bolstering their own Persian Gulf and Indian Ocean capabilities, or by some combination thereof.

It would be untrue to say that Western Europe and Japan did not share America's concern with the Soviet threat. According to CIA analyst Frans Bax, "The Soviet invasion of Afghanistan was certainly as much of a shock in other Western capitals as it was in Washington."[42] Indeed, in May 1980 NATO's Defense Planning Committee issued a communiqué which read in part

> Ministers agreed that the stablity of regions outside NATO boundaries, particularly in the South West Asia area and the secure supply of essential commodities from this area, are of critical importance.[43]

The following month an even stronger critique of the Soviet invasion was expressed by the NATO ministers who were meeting in Ankara.[44]

But collective words did not readily translate into common action. The problem, once again, was that the allies placed different weights on their various political and economic interests. They also had different views regarding the appropriate tactics for realizing these interests. For the European allies and for Japan, access to energy sources predominated other regional problems. Accordingly, the allies dissociated themselves from U.S. policy regarding the Middle East in general and Afghanistan in particular. The European community continued to promote the Euro–Arab dialogue, while many individual countries pursued arms-for-oil deals which Washington criticized as being destabilizing. Further, although several countries bolstered their naval forces in the Indian Ocean area, all except Britain refused to participate in joint maneuvers with U.S. vessels.

The alliance response to Afghanistan was further complicated by the growing energy ties between the Soviet Union and both Europe and Japan. Ironically, after the 1973 oil embargo the United States itself had encouraged the development of Soviet oil and gas fields. In 1978, Soviet and European negotiators began to hold informal discussions regarding the construction of a natural gas pipeline from Siberia to Western Europe. In June 1980, at a time when NATO ministers had joined in criticism of the Soviet invasion of Afghanistan, German chancellor Helmut Schmidt visited Moscow to launch official talks. Curiously, President Carter made no effort to block these East–West energy deals. According to Defense Department official Sumner Benson, "Some Carter officials joined their European counterparts in arguing that development of Soviet oil and gas reserves would benefit both East and West."[45] With the Central Intelligence Agency warning that the Soviets were running out of easily recoverable domestic resources, it seemed prudent to help Moscow develop new supplies. The alternative might be a sudden grab for Persian Gulf reserves, now that the Red Army was nearby.

A collective alliance response to the Soviet invasion of Afghanistan was therefore blunted owing to conflicting views over how to respond to Moscow, growing European interest in Soviet energy resources, doubts about the efficacy of U.S. military power in securing Middle East oil fields, and European fears of entanglement in an East–West conflict. At the macropolitical level, the allies agreed that the Soviet Union represented a threat to their national security. But at the tactical level, divergent views regarding strategy were found. To the allies, détente and East–West energy trade promoted rather than diminished their security; with regard to the Persian Gulf, diplomacy rather than military force was the appropriate tool of statecraft.

The alliance response to Afghanistan did not bode well for a collective response to a future energy crisis. It now appeared that the United States might respond to another shock through the use of military force, a tactic that could make energy matters worse. This, in combination with the growing political instability of the Middle East, placed a premium on national energy security policies, including stockpiling, conservation, fuel switching, and diversification of suppliers. Indeed, during the first quarter of 1980, following the Soviet invasion south, stockpiles in OECD countries reached record levels. No less significant was the fact that the United States cut its demand for oil imports by over 1 percent, and U.S. oil consumption for the year was well below the ceiling set during the Tokyo summit. Ironically, the Iranian revolution and invasion of Afghanistan may have promoted alliance energy security after all.[46]

The Iran–Iraq War

The IEA's governing board met several times in early 1980 to reconsider its actions (or lack thereof) during the Iranian revolution and to consider modifications to its operations. In January 1980 the energy ministers agreed to increase their emergency oil stockpiles to 90 days of net oil imports. As had been apparent since the IEA opened its doors, stockpile management was becoming the focal point of national and alliance energy security programs. Stockpiles, if properly managed, not only served to buffer the initial shock caused by an oil shortage, but they also served to deter OAPEC from declaring another embargo.[47]

Stockpiles also represented a realistic response to the growing problems inherent in the emergency allocation system. By 1980, over 40 percent of OPEC oil was moving to customers in the form of direct deals; the traditional flexibility of the international oil companies was being undermined.[48] In a future crisis, IEA recognized that the ability of OPEC to exercise wide-ranging destination restrictions would complicate the allocation process. As a result, doubts were cast on the effectiveness of the agency's emergency procedures; in April the IEA's godfather, Henry Kissinger, declared that the allocation system was "inadequate."[49]

In May 1980 the governing board began to confront the problem of how the agency should respond to oil shortfalls of less than 7 percent, as was the case during the Iranian crisis. While no specific agreement on intervention was reached, the ministers determined that they would meet quickly if "tight market conditions appear imminent." They also agreed to consult on stockpile releases and oil import policies. These were hardly the sharp teeth that some observers hoped the IEA would grow, but they indicated that the agency was disturbed by its inability to act during the Iranian shortfall, with the threat to its legitimacy that failure implied. If nothing else, the agency could now create a mirage of alliance cohesion during the early days of a crisis by calling together the member states' energy ministers, something the allies had failed to do since 1973.[50]

The IEA was soon given the opportunity to prove its alliance utility. In September 1980, territorial and ideological conflicts between Iraq and Iran boiled into open hostility. During the early days of the fighting, Iraq took the offensive, and by November it had penetrated 80 kilometers into Iran, besieging the refinery center at Abadan. During the ensuing Iranian counterattack, Iraqi oil installations and pipelines were destroyed. The conflict removed some 3.8 million barrels of oil per day from world markets.[51]

As promised, the IEA governing board met in mid-September to analyze the oil situation and formulate an appropriate policy response. Although IEA stockpiles were at record high levels, once again uncertainty about the size and duration of the shortages led to an initial surge of spot market prices, with prices leaping to $40 per barrel. On October 2 the board determined, however, that stockpile withdrawals must provide the first line of defense against panic and price shocks. At the same time, IEA members agreed to prevent oil firms from making "abnormal purchases" on the spot market. Nonetheless, the Japanese pursued the direct-deal route for additional supplies, signing a contract with Kuwait for oil that included a $5 per barrel premium.[52]

In December the ministers reconvened, basically reaffirming the tack previously taken. The secretariat was also instructed to continue the monitoring of supplies and identify any severe imbalances that might be occurring within the IEA territory. Governments were asked to support the oil companies in seeking ad hoc solutions to any imbalances that were identified. The board further stressed the importance of stock drawdowns, in order to reduce net oil imports by 10 percent during the first quarter of 1981.[53]

At least one IEA member, Turkey, complained that the informal policies adopted by the agency were inadequate. In 1980 Turkey was completely exposed to an oil shock. Iran and Iraq had been its major crude suppliers, providing 60 percent of its supplies, and now both exporters were down. Its stock position was low and quickly deteriorating. Its relations with the international oil companies was poor, owing to some outstanding bills that had gone unpaid. During the fourth quarter of 1980, Turkey asked the IEA secretariat for help in meeting its oil shortage.

Consultations between the IEA secretariat, the United States, and the oil companies followed this plea for help. Unfortunately for the agency's credibility, these talks dragged, as Turkey balked at the price being offered for cargoes and the United States failed to provide extraordinary assistance. According to former IEA official Daniel Badger, "Nothing was . . . done to help Turkey before Christmas, at which point internal shortages were widespread and stocks reported to be down to between 5 days and zero." A desperate situation was only saved in January by the reopening of an Iraqi pipeline.[54] In a review of the IEA's response to the Iran–Iraq war, a Department of Energy study declared that the agency had "failed to fulfill its promise."[55]

By 1981 any fears of a war-induced energy crisis had passed. Saudi Arabia had bolstered its production, partly to thank Washington for American military protection against Iran (American-manned AWACS aircraft were patrolling Gulf airspace) and partly in the hope of obtaining advanced weaponry.[56] Iran and Iraq had begun exporting oil again, albeit in limited quantities. Oil companies refrained from panic buying for stockpiles, as high interest rates and sluggish demand in recessionary economies discouraged investment in working capital. As a result of these factors, spot prices for oil declined during the first quarter of 1981, and no official price hikes were decreed by OPEC after its unilateral increase of December 1980.

Despite shortfalls of similar magnitude, the Iran–Iraq war caused far less disturbance on international energy markets than the Iranian revolution of 1978–1979. Various factors help to explain this outcome, including supply/demand conditions, macroeconomic variables, and stockpile levels. But the IEA's role, even if marginal, should not be overlooked. By providing information on market conditions and a meeting place for Western energy ministers, it contributed to the formulation of an appropriate energy policy response.[57]

Conclusion

The International Energy Agency was created in 1974 to demonstrate that the Western alliance had not died during the Arab oil embargo. By encouraging member states to adopt crisis prevention measures, the agency was meant to deter potential oil blackmailers from launching embargoes. While the causality cannot be determined, the agency may take heart from the fact that no embargoes *have* occurred since it started. But its record in preventing energy scrambles has been less noteworthy.

During the Iranian revolution the agency failed completely to promote alliance cohesion. A small oil shortfall was translated into a huge Western panic as states rushed to the spot market and directly to producers. The IEA was a helpless bystander, and its coordination attempts occurred late in the day and without authority. The member states went the unilateral route, and as a result every state was forced to pay higher prices for oil. In this the United States was no different from any other Western ally, acting "small" rather than as a leader.

Yet the United States and other IEA members apparently

learned something from the Iran imbroglio and the subsequent Soviet invasion of Afghanistan. President Carter became a reborn Cold Warrior, recognizing the utility of military power in a dangerous world. On the energy front, the United States and its allies built up their stockpiles as a first line of defense against future crises. In the absence of a supplier of last resort, the IEA members recognized that crisis prevention and national measures must be stressed. By providing market information and a pool of expert advice, the IEA helped alliance members to formulate appropriate energy policies. These policies contributed to the relatively successful weathering of the 3 million barrel per day shortfall that accompanied the outbreak of the Iran–Iraq war.

By the early 1980s it was clear that the allies faced three prominent threats to their energy security: Islamic fundamentalism, regional conflict in the Middle East, and the external Soviet threat. The United States, despite French pretentions, was the only alliance actor that could hope to contain these threats and maintain the Middle East oil trade. In developing its special relationship with Saudi Arabia, the United States helped to win higher output levels in the autumn of 1980. In working with the Soviets, it helped to contain the Iran–Iraq war. In ultimately organizing tanker convoys through the Persian Gulf, it preserved the West's freedom of navigation. U.S. action during the Iran–Iraq war, coupled with responsible energy policies that led to a drop in demand and high stockpile levels, suggested a reassertion of its leadership over the Western alliance.

To be sure, it would be hard to argue that this resurgence of American power can be easily converted into collective action during a future energy crisis since the United States is still in no position to serve as the allies' emergency supplier. The IEA offers Western officials the opportunity to cooperate in overcoming shortages; whether or not they actually do so depends on state interests. But so long as the allies share a collective set of interests during a crisis, attempts to coordinate policy can be expected, even in the absence of a hegemonic distribution of power. These coordination attempts may not be so effective as those which occurred when the United States served as energy supplier of last resort (some cheating on agreements is probable), but it is better than no coordination at all.

9

Conclusions

Economic means are useful, even indispensible, to . . . cement a coalition.

RAYMOND ARON[1]

At the end of World War II the United States faced an international political economy that was inhospitable to the spread of its liberal goals. Continental Europe verged on economic chaos, providing fertile ground for social unrest and the adoption of alternative philosophies. The Soviet Union controlled Eastern Europe, while Britain maintained a sterling bloc of Commonwealth nations. American officials recognized that without decisive leadership on their part, liberation might only bring with it misery, strife, and further international conflict.

U.S. decision makers had no doubt that economics and security were inextricably linked in the postwar world. Europe confronted severe shortages of food and fuel, and these had to be overcome before grandiose international schemes were emplaced. Indeed, the economic crises of the postwar era gave the United States an opportunity to assert its leadership, and to gain the legitimacy that was required before a cooperative liberal order could be built.

The Cold War that emerged soon after the war's end derailed any idealistic visions of a truly international economy. But members of the new Western alliance recognized that cooperation must go beyond the politico-military sphere and incorporate mutual economic problems as well. Recovery and rearmament demanded, among other things, a vigorous industrial base, funds for investment, and adequate supplies of food, energy, and strategic minerals.[2]

I have argued in this book that energy security—meaning assured access to fuel at reasonable cost—has been a prominent concern of the Western allies throughout the postwar period. Alliance policymakers have viewed energy security as vital to economic strength, political stability, and military preparedness. Energy crises could undermine and weaken the alliance, as states scrambled for available supplies, and as national governments reeled under the economic and strategic shockwaves that crises produced. The potential reverberations of an energy shortage were sharply drawn by State Department economist Charles Kindleberger in 1946 when he said (as cited in Chapter 2), "France will go communist if the demands of the French for coal . . . are not met."

This concluding chapter, like a Sunday sermon, has three objectives. First, I assess alliance relations during periods of energy shortage, using the conceptual framework developed in the opening chapter. Second, I offer some reasons as to why alliance energy security has proved to be an elusive goal. Finally, the book's findings are placed in terms of the relevant international relations literature, and suggestions for future research are provided.

History and Theory

I have argued that alliance relations during energy crises are best understood in terms of a conceptual framework that incorporates hegemonic power on the one hand, and collective interests on the other. Hegemonic power, I show, has been a sufficient condition for the generation of alliance policy *coordination,* but not *cooperation.* The tenor of alliance relations has been determined by the strategic interests of the member states; at certain times the allies attached similar priorities to their vital interests, while at other times they weighed their interests quite differently.

When interests were shared, and hegemonic power was present, the allies engaged in mutually beneficial policy coordination; in other words, they cooperated. Examples are provided by the postwar coal crisis, when the European Coal Organization was founded, and by the alliance response to Iran's nationalization of British oil assets in 1951. At the war's end, the coal crisis stood in the way of economic rehabilitation, and it threatened the political stability of fragile western governments. A scramble for available supplies would have undermined European economies further, and it raised the specter of renewed conflict. The United States feared that the coal shortage could sabotage its ambitious postwar plans for the international economy. As a result, it worked with its western allies to overcome the shortfall, and led in the creation of a multilateral institution. The fact that the Soviet Union was invited to join the European Coal Organization but refused to do so should not be forgotten. For immediate economic reasons—and perhaps for far-sighted strategic ones as well—Stalin had no interest in sharing coal supplies with his former allies.

In Iran, the tactical differences between the United States and Great Britain over how to handle the crisis might lead us to conclude that fundamental interests were in conflict, but this is not the case. The alliance division of labor called for Great Britain to defend the Middle East in the event of a world war with the Soviet Union. The United States was in no position to supplant Britain at the time, and thus it had no choice but to provide political, economic, and ultimately covert intelligence support for its most important ally. Hegemonic power was placed in service of a common strategic purpose.

When interests were in conflict and power resources were present, the hegemon coerced alliance policy coordination. The most notable case is provided by the Suez crisis of 1956, although a similar theme emerged a decade later during the Six Day War. At Suez, the United States had a fundamental interest in preventing Soviet advances in the Middle East. In contrast, the British and French (and the Israelis) were determined to depose Nasser, with apparently little regard for the global consequences of their actions. The allies were not only abandoned by their leader, but the United States used coercive measures to force their withdrawal from the canal zone. Washington refused to extend energy security to its allies when their strategic interests conflicted with its own.

The lines were less sharply drawn in June 1967, but once again alliance interests differed. The United States provided Israel (which itself had been abandoned by such former allies as France) with diplomatic and military backing, and it sought to contain a Soviet Union that now had several Middle East clients. The Western allies, out of concern for their energy security, dissociated themselves from U.S. policy, and they initially refused to mobilize the emergency plans of the OECD oil committee, fearing they might antagonize the Arab producers. Only after an implicit threat on the part of the U.S. delegate did the committee find that the "threat of an emergency" existed.

When the alliance was characterized neither by a hegemonic distribution of resources nor by a set of shared interests, the outcome was discord and conflict, as exemplified by the Arab oil embargo of 1973–1974. The allies scrambled for oil, realizing the nightmare that many Western officials had long feared. Fifteen years after the event, it is easy to minimize the economic and strategic damage that the energy crisis wrought on the Western alliance, but the oil shock weakened alliance economies, caused mistrust among member states, and worked only to the advantage of parties hostile to the West.

A more ambiguous picture is presented by the Iranian revolution of 1978–1979, and the Iran–Iraq war of 1980. In these cases, the allies had a set of collective interests, but hegemonic power was absent. I argued in Chapter 1 that *attempted coordination* would be the expected outcome. This occurred at both times, but in the first case it was of no consequence, while in 1980 some marginal benefits may have accrued. These differences, it must be admitted, cannot be explained adequately by my sparse framework. The framework does help us to understand, however, why the International Energy Agency remains in existence. So long as the allies maintain shared interests in the energy area, attempts at policy coordination will be an enduring feature of their collective life.

Our concern with both power and interests in alliance relations helps us to see some of the important limitations associated with basic hegemonic stability theory. To begin with, the theory does not adequately explain why the hegemon has used positive sanctions at some times, and negative sanctions at others; it does not tell us much about the different ways in which alliance policy coordination may be achieved. It dichotomizes the problem of coordination by focusing exclusively on "cooperation" and "conflict," as if these categories exhausted the possibilities. Parsimony is a desirable feature in any theory, but it should not obscure or distort the historical record.

Second, in positing the reasons for hegemonic decline, theorists who concentrate on power resources must abandon the systemic level of analysis for the unit or state level. For example, Robert Keohane argues that the fall of American energy power was brought about by "special interests" who "drained America first" and "prevented the implementation of a farsighted strategic policy of coordination . . ."[3] In short, hegemonic decline is caused by domestic actors.

In this study I have certainly not overlooked the importance of particular interests in U.S. foreign policy execution; to the contrary, the role of such domestic actors as coal miners and oil producers has been highlighted. But I have suggested that friction within the alliance itself, caused by the conflicting strategic interests of the member states, has been a significant factor in hegemonic decline. After Suez, Western Europe vigorously pursued the development of new energy sources, including nuclear power, and diversification of suppliers, such as Algeria and the Soviet Union. The events at Suez also prompted the French to build an independent nuclear deterrent, equating power in the modern world with the possession of atomic weapons. After 1956, the allies realized that they could not depend on the United States to provide for their economic and military requirements in every scenario; to their surprise, Washington had abandoned and coerced them when they were most vulnerable. In sum, the allies purposely sought to increase their scope for independent action in the international system. To them, American hegemony had revealed a malign side.

Searching for Energy Security

Reflecting on the seven postwar energy crises, the reader may wonder why the allies have proved incapable of ensuring their energy security. On its face, the problem of energy security does not appear to be an intractable one. By stockpiling fuel and by diversifying sources and suppliers, it would seem that this objective could be achieved.

But the energy security problem cannot be viewed apart from the broader panorama of alliance relations in which it has been embedded. Between World War II and 1956, the allies relied on the United States to serve as energy supplier of last resort; it was impossible to imagine a situation in which Washington would purposely withhold fuel. The United States had fashioned a hemispheric energy policy during the Marshall Plan years in which western hemisphere resources were conserved in part to meet alliance contingencies. So long as the United States generously maintained spare production capacity, the allies had no incentive to pursue energy security measures on their own. In effect, they could act as "free riders" on America's oil reserves.

After Suez, the quest for alliance energy security grew more complex. For their part, the European allies wanted to retain access to U.S. resources in the event of an *alliance* crisis, but some of them also wanted to enjoy a sufficient degree of energy independence in order to pursue their unilateral foreign policy interests. Accordingly, they developed new energy sources, and sought alternative suppliers, including the Soviet Union. In contrast, Washington wanted to maintain allies who were strong yet dependent. Nuclear power was attractive to Washington for this reason, since the United States controlled the production of a critical input, enriched uranium. Despite all the energy policy activity in the OEEC and OECD oil committees, the gradual erosion of hegemonic power, and doubts about alliance strategic purpose, made the post-Suez search for energy security an elusive goal.

Curiously, the European allies did little during the 1950s and 1960s to build their strategic stockpiles of oil, even though this would appear to have been their cheapest and easiest pathway to energy security. Stockpile build-up had long been advocated by economists and energy policy experts; indeed, it was mandated by the OEEC oil committee. Why was this option not pursued more vigorously?

Several answers present themselves. First, the allies feared that stockpiling of oil would be regarded as a hostile act by oil-producing states, prompting the development of a producers' cartel. Second, stockpile releases during a shortage had the character of a public good. If one state released oil, it would lower prices for all consumers, whether or not they had adopted similar measures. As economist John Weyant has written, "because the benefits of stock releases are widely shared and the costs are imposed only on those who hold the stocks, there is a disincentive for individual countries to hold stocks if they believe other countries will do so."[4] Finally, stockpiling programs meant an increase in oil imports for nonproductive purposes, exacerbating already severe balance-of-payments problems.[5] For all these reasons, stockpiles were an underutilized weapon in the post-Suez energy security arsenal.

Accordingly, the allies' search for security focused on alternative solutions, preferably payable in "soft" national currencies, including the maintenance of high-cost coal mines, the development of nuclear power, and exploration of oil in "colonial" areas where national currencies could be used. None of these approaches were successful. As discussed in Chapters 2 and 3, Western Europe's coal mines after World War II faced severe labor and capital problems. Despite numerous national programs to rehabilitate the mine sector, and substantial Marshall Plan aid, the high price of European coal discouraged its utilization. Economist M.A. Adelman calculated that the cheapest grade of coal in the European Coal and Steel Community was priced at $16 per metric ton at the mine in 1958. Given the difference in heat value per ton (oil has more heat value than coal) coal would have been competitive with fuel oil only if oil had cost $21 per ton. But oil was available in Western Europe at the time for about $12. This $9 "security premium" was more than European countries were willing to pay for domestic coal.[6]

Additionally, there was a competitive element that impeded use of coal. For example, if the Germans fueled their industries with high-priced domestic coal, while the French exploited cheap Saharan oil, German industries would be placed at a competitive disadvantage. A trade-off existed between international competitiveness and energy security, which could have only been solved by complete energy policy coordination.[7]

Oil diversification strategies also failed to provide energy security. Despite its proximity to western Europe, North Africa was no more politically stable than the Persian Gulf or Near East. Algerian "terrorists" could easily cut pipelines and disrupt production. New governments could nationalize or expropriate facilities. Further, the various oil-producing countries could form a cartel, which of course they eventually did in 1960. To be sure, North African oil did not have to transit the Suez Canal, but this proved to be small comfort in light of many other destabilizing factors.

The Soviet Union became another alternative supplier after the Suez crisis. But growing dependence on Soviet oil created intra-alliance problems, as the United States discouraged western European countries from aiding in East–West pipeline construction and bilateral barter deals. The prudent level of alliance dependence on Soviet-controlled energy sources had been a point of dispute since the postwar coal crisis; it endured without resolution through the 1980s, when the allies conflicted over the building of a natural gas pipeline from Siberia to the Western European border. Indeed, the pipeline dispute provides an additional test of our conceptual framework. With interests in conflict, and U.S. hegemonic power lacking, the outcome of the dispute was alliance discord.

Yet an additional reason may be cited for the failure of the allies to ensure their energy security. As we have seen, for some states oil shortages have provided crises of opportunity. The French, and, to some extent, the Japanese, saw in the Arab oil embargo of 1973–1974 an opportunity to break the detested Anglo-Saxon oil monopoly in the Middle East, and to advance national interests. So long as tension existed between collective alliance and individual national interests, it would prove impossible to formulate effective energy security strategies.

This brings up one final point. During the early 1970s, most analysts perceived that the Arab–Israeli conflict was the predominant if not sole threat to alliance energy security. But energy crises have arisen from numerous sources, including nationalism and various regional conflicts. Were it true that the energy problem was localized within a single political conflict, it may have been easier to manage. The search for collective energy security, however, has been made much more complex owing to the diversity of crisis scenarios and, consequently, the variety of alliance interests at stake.

Nonetheless, with the establishment of the International Energy

Agency, the allies reaffirmed their collective interest in supply security, and they recognized that this goal could not be readily achieved without assistance from the United States. But the IEA also made manifest the degree to which the postwar energy security structure had come apart. In the years following the Arab oil embargo, coordination of *national* policies—rather than hegemonic management—became the basis for collective security. Every state would, in the first instance, have to ensure its energy security on its own. The United States could no longer be relied upon to meet emergency needs, unless the alliance was at war. The allies faced the challenge of policy coordination in the absence of a hegemonic power.[8]

It should be emphasized that the shift away from hegemonic management of energy crises has been made explicit in recent U.S. policy statements. In the comprehensive report *Energy Security,* published by the Department of Energy in March 1987, the focal point of the section on "international energy actions" was placed squarely on strategic stockpiles. "The United States," the report asserted, "has been encouraging its partners in the International Energy Agency to undertake a political commitment to increase emergency stocks and to establish a consultative process for drawing down such stocks in a coordinated fashion in the event of an oil supply disruption."

The DOE recommended that, in future crises, early and regular consultation should occur, rapid exchanges of information must take place, and IEA members should monitor *one another* (my emphasis) as to their participation in a coordinated stockpile drawdown. The report nowhere suggests that IEA members can expect to receive emergency shipments of U.S. oil supplies in future.[9] Perhaps without knowing it, the authors of *Energy Security* confirmed the 1970 analysis of the Cabinet Task Force on Oil Imports, which had first warned that the decline in U.S. spare production capacity meant that Texas could no longer be counted upon to overcome alliance shortfalls. The West's response to that decline has been slow, and uncertainty remains about the effectiveness of the IEA's emergency management system.

In addition to stockpiling, the United States, and many of the western allies, have also adopted market-oriented energy security strategies since the 1970s. The price shocks of that decade, when allowed to ripple through the economy without government interference, prompted enormous changes in energy supply and demand patterns. A significant amount of energy resource development occurred in western Europe itself, with exploitation of the North Sea and the Dutch gas fields being the most prominent examples. At the same time, capital expenditures were made by western industries in fuel-efficient technologies, and automobile fleets became more economical.

These trends appeared to enhance alliance energy security, and leave the oil cartel in disarray, at least in the short-run. Over the longer-term, however, the free-market approach may have ironic consequences. The Arab countries remain the low-cost producers of oil, and Western dependence on OAPEC is once again rising.

Thus, despite the complacency of the 1980s, energy insecurity is likely to remain a chronic ailment of the Western alliance. This is not to detract from the International Energy Agency; to the extent that the IEA has helped states coordinate their national policies, and made energy markets more transparent, it has promoted collective alliance goals. But given the absence of hegemonic power on the one hand, and divergent alliance interests in the Middle East (and elsewhere) on the other, the alliance cannot be said to have solved its energy problems.

Managing Alliance Relations

Michael Ward once wrote that "the study of alliances has been relatively uninspiring."[10] In general, it has not been characterized by cumulative research. A premium has been placed on abstract theory-building, to the detriment of empirical tests.

This complaint about work in political science is not a new one. Writing in 1968, Oran Young urged scholars of international relations to upgrade "empirical analysis relative to conceptual work" and to "combine the two activities in fruitful ways." He implored us "to match concepts and realities."[11] Twenty years later, this call remains urgent.

Alliance relations provide an extraordinary field in which to conduct basic research. Despite the centrality of alliances in international political theory, we still have relatively few works that detail their creation, and even fewer about their maintenance.[12] Further, the literature has focused almost exclusively on politico–military aspects of alliances; economic issues have been largely untouched. With renewed debate over such issues as defense burden-sharing and East–West trade, the economic dimension of alliances deserves sustained analysis. The quote from Raymond Aron that opens this chapter is suggestive of the role economic factors play in coalition maintenance, but detailed tests of his hypothesis are lacking.

In this book, I have argued that alliance security requires more than a military agreement on how to meet a common threat. Security also depends upon appropriate economic policies that ensure members states with the resources needed to support their defense industrial base. In the absence of such policies, alliance cohesion may be strained and economic vitality sapped as states fall victim to competitive scrambles for scarce resources. The issue is not necessarily whether the alliance will *break* as a result of energy and economic crises; it is the extent to which alliance power is weakened vis-à-vis the adversary.

To be sure, as Hedley Bull noted shortly before his death, the Western alliance has managed to endure many economic and strategic disputes, from the Suez crisis of 1956 to the Arab oil embargo of 1973. He stated that "differences of interest among their members should not be regarded as fatal to alliances, which never rest upon a complete identity of interests and only require a partical one."[13] But the alliance has changed in significant ways since Suez. Prior to 1970, the United States could exercise overwhelming leverage in almost every issue-area. Bull himself recognized that it was Washington's "preponderance of power—economic and political as well as military" that gave it a "privileged position in Alliance decision making."[14]

Today, the United States faces a more difficult task in promoting Western policy coordination. Not only has its hegemonic position in energy eroded—in part because of alliance frictions—but its grip over the international political economy has weakened in the face of a resurgent western Europe and a world-beating Japan. American diplomacy at the Washington Energy Conference of 1974 provides a telling example. In order to win an alliance agreement on creation of the IEA, President Nixon had to threaten the withdrawal of American troops from Western Europe; he had to link the energy area, in which the United States had become weak, to the military one, where it remained strong. This strategy reflected America's relative decline in specific sectors as much as its underlying capabilities.

Yet to analyze the western alliance simply in terms of the distribution of power resources would be a mistake.[15] After all, its durability also rests upon the interests of the member states. The challenge for the industrial democracies today, as in the past, is to formulate collective responses to shared economic and security concerns. Alliance cooperation may prove difficult to achieve when hegemony is absent, but without common strategic purpose it is impossible.

Notes

Chapter 1

1. Michael Taylor, "The Theory of Collective Choice," in Fred Greenstein and Nelson Polsby, eds., *Handbook of Political Science*, v. 3, (Reading, Mass.: Addison-Wesley, 1975), p. 413.

2. Glenn Snyder and Paul Diesing, *Conflict Among Nations* (Princeton, N.J.: Princeton University Press, 1977), p. 23.

3. Harry Eckstein, "Case Study and Theory in Political Science," in Fred Greenstein and Nelson Polsby, eds., *Handbook of Political Science*, v. 7, (Reading, Mass.: Addison-Wesley, 1975), p. 104.

4. See Duncan Snidal, "The Limits of Hegemonic Stability Theory," *International Organization 39* (Autumn 1985): 579–614.

5. See Charles P. Kindleberger, *The World in Depression: 1929–1939* (Berkeley: University of California Press, 1973); Robert Gilpin, *War and Change in World Politics* (New York: Cambridge University Press, 1981). For an analysis that argues the possibility of posthegemonic cooperation, see Robert Keohane, *After Hegemony* (Princeton, N.J.: Princeton University Press, 1984).

6. Keohane, *After Hegemony*, p. 32.

7. Robert Gilpin, *U.S. Power and the Multinational Corporation* (New York: Basic Books, 1975), pp. 103–104.

8. For more on this methodological point, see Eckstein, "Case Studies."

9. Snyder and Diesing, *Conflict Among Nations*, p. 183.

10. George Liska, *Nations in Alliance* (Baltimore: Johns Hopkins University Press, 1962), p. 12.

11. Hans Morgenthau, "Alliances in Theory and Practice," in Arnold Wolfers, ed., *Alliance Policy in the Cold War* (Baltimore: Johns Hopkins University Press, 1959), p. 185.

12. Michael Ward, "Research Gaps in Alliance Dynamics," *Monograph Series in World Affairs 19* (Book 1, 1982).

13. Kenneth Waltz, *Theory of International Politics* (Reading, Mass.: Addison-Wesley, 1979), p. 167.

14. See James Schlesinger, *The Political Economy of National Defense* (New York: Praeger, 1960).

15. The treaty may be found in NATO, *The North Atlantic Treaty Organization* (Paris, 1957).

16. See David Deese, "Oil, War, and Grand Strategy," *Orbis* (Fall 1981), p. 555; N. J. Spykman, *America's Strategy in World Politics* (New York: Harcourt, Brace, 1942), p. 97; Hans Morgenthau, *Politics Among Nations* (New York: Knopf, 1968), pp. 112–113; Schlesinger, *Political Economy of National Security*.

17. See Donald Goldstein, "Energy as a National Security Issue," in Donald Goldstein, ed., *Energy and National Security* (Washington: National Defense University Press, 1981), p. 7; Bruce Russett, "Security and the Resources Scramble: Will 1984 Be like 1914?," *International Affairs 58* (Winter 1981–1982), p. 43; Peter Cowhey, *The Problems of Plenty* (Berkeley: University of California Press, 1985), p. 37.

18. See Dennis Mueller, *Public Choice* (Cambridge, U.K.: Cambridge University Press, 1979). Robert Keohane also makes this argument in *After Hegemony*.

19. See Oran Young, *The Politics of Force* (Princeton, N.J.: Princeton University Press, 1968), pp. vii–viii.

20. Snyder and Diesing, *Conflict among Nations*, pp. 129–152.

21. Glenn H. Snyder, "The Security Dilemma in Alliance Politics," *World Politics 36* (July 1984): 461–495.

22. See Charles A. Kupchan, *The Persian Gulf and the West* (Boston: Allen & Unwin, 1987).

23. Thomas Schelling, *Arms and Influence* (New Haven, Conn.: Yale University Press, 1966), p. 97.

24. Young, *Politics of Force*, p. 34.

25. Snyder and Diesing, *Conflict Among Nations*, p. 8.

26. Robert Keohane, "The Demand for International Regimes," in Stephen D. Krasner, ed., *International Regimes* (Ithaca, N.Y.: Cornell University Press, 1983), p. 150.

27. Ibid.

28. Young, *Politics of Force*, p. 394.

29. Kupchan, *Persian Gulf and the West*, p. 181.

30. Raymond Aron, *Peace and War* (Garden City, N.Y.: Doubleday, 1966), p. 62.

31. Joanne Gowa, "Hegemons, IOs, and Markets: The Case of the Substitution Account," *International Organization 38* (Autumn 1984), p. 663.

32. Keohane, *After Hegemony*, p. 140.

33. Stephen Krasner, "A Statist Interpretation of American Oil Policy toward the Middle East," *Political Science Quarterly 94* (Spring 1979), p. 82.

34. See especially Stephen D. Krasner, *Defending the National Interest* (Princeton, N.J.: Princeton University Press, 1978), and Stephen D. Krasner, "Domestic Constraints on International Economic Leverage," in Klaus Knorr and Frank Trager, eds., *Economic Issues and National Security* (Lawrence, Kans.: Regents Press of Kansas, 1977).

35. Krasner, "A Statist Interpretation," p. 79.

36. Krasner, *Defending the National Interest*, p. 61.

37. Ibid., p. 74. On corporate power see also Charles Lindblom, *Politics and Markets* (New York: Basic Books, 1977); Carl Kaysen, "The Corporation: How Much Power? What Scope?," in Edward S. Mason, ed., *The Corporation in Modern Society* (Cambridge, Mass.: Harvard University Press, 1959).

38. David Painter, *Oil and the American Century* (Baltimore: Johns Hopkins University Press, 1986).

Chapter 2

1. U.S. War Department, *The European Coal Situation*, May 11, 1945, Record Group (RG) 260, Records of the War Department, box 103.

2. U.S. Bureau of the Budget, "Mission for Economic Affairs," November 1944, Records of Allied Operational and Occupation Headquarters, RG 331, box 87.

3. On UNRRA see Robert Pollard, *Economic Security and the Origins of the Cold War* (New York: Columbia University Press, 1985), pp. 26–30.

4. Hull to Stimson, quoted in Thayer to Scowden, July 13, 1944, RG 331, box 87.

5. Cordell Hull, *Memoirs* (New York: Macmillan, 1948), p. 81.

6. Quoted in Lloyd Gardner, *Economic Aspects of New Deal Diplomacy* (Madison: University of Wisconsin Press, 1964), p. 304.

7. Pollard, *Economic Security*, p. 32; Gardner, *Economic Aspects of New Deal Diplomacy*, pp. 304–325.

8. M.I.T. Center for International Studies, *The Objectives of United States Economic Assistance Programs* (Washington, Government Printing Office, 1957).

9. N. J. D. Lucas, *Energy and the European Communities* (London: Europa, 1977), p. 1.

10. Walter Thayer to War Department, July 3, 1944, Records of the War Department, General and Special Staffs, Civil Affairs Division, RG 165, box 167.

11. Thayer to Scowden, July 13, 1944, RG 331, box 87.

12. Mission for Economic Affairs (MEA), "The European Coal Problem in the Immediate Postwar Years," August 22, 1944, Records of the Combined Boards, RG 179, box 7.

13. European Coal Organization (ECO), *The European Coal Organization* (London: European Coal Organization, 1947).

14. F. S. V. Donnison, *Civil Affairs and Military Government* (London: HMSO, 1961), pp. 395–396.

15. Donnison, *Civil Affairs*, p. 395; interview with Ambassador Samuel Berger, Washington, September 26, 1979; interview with the Honorable Nathaniel Samuels, New York, April 8, 1985 (Samuels was a U.S. delegate to the ECO).

16. Quoted in William Chanler to James Dunn, October 26, 1944, 840.6362/10-26-44, RG 59; Chanler was in the War Department's Civil Affairs Division; Dunn in State's Office of European Affairs.

17. SHAEF, Solid Fuels Section, "Semi-Annual Report," January 9, 1945, RG 331, file 319.1, box 87.

18. James Riddleberger, interview, Harry S. Truman Library (HSTL) oral history collection.

19. Jacques Freymond, *The Saar Conflict* (New York: Praeger, 1960); A. W. DePorte, *De Gaulle's Foreign Policy: 1944–1946* (Cambridge, Mass.: Harvard University Press, 1968).

20. Office of European Affairs, "Memorandum of Meeting Between Mr. Byrnes and M. Bidault," August 24, 1945, Records of the Office of European Affairs, RG 59, box 14A.

21. For a detailed account of the German recovery question see John Gimbel, *The Origins of the Marshall Plan* (Stanford, Calif.: Stanford University Press, 1975).

22. Winant to War Areas Economic Division, March 17, 1945, 840.6362/3–17–45, RG 59.

23. Grew to London Embassy, February 21, 1945, 840.50/2–21–45, RG 59.

24. Gaddis, *Strategies of Containment*, pp. 5–7.

25. Harriman to Secretary of State, March 4, 1945, 840.50/4-4-45, RG 59.

26. Harriman to Secretary of State, March 21, 1945, 840.50/3-21-45, RG 59.

27. Ambassador Samuel Berger, interview, Washington, September 26, 1979.

28. Charles P. Kindleberger, "Toward the Marshall Plan: A Memoir of Policy Development, 1945–1947," mimeograph, December 1985, p. 13.

29. On the history of the ECO see Nathaniel Samuels, "The European Coal Organization," *Foreign Affairs 26* (October 1948): 728–736.

30. ECO, *European Coal Organization*.

31. Donnison, *Civil Affairs*, p. 413.

32. R. M. Blount to Harry Truman, August 1, 1945, 840.6362/8-3-45, RG 59.

33. ECO, *European Coal Organization*.

34. SHAEF, Solid Fuels Division, "An Evaluation of the Coal Situation in Northern Europe," May 18, 1945, RG 331, box 183.

35. Ibid.

36. C. J. Potter and Lord Hyndley, "Report of the Potter/Hyndley Mission to Northwest Europe," June 19, 1945, paper 60, RG 179; also in *Foreign Relations of the United States (FRUS): 1945*.

37. Truman to De Gaulle and Churchill, June 24, 1945, 840.6362/6-24-45, RG 59.

38. Clay to McCloy, June 29, 1945, in Jean Edward Smith, ed., *The Papers of General Lucius D. Clay, Germany: 1945–1949*, v.1, (Bloomington: Indiana University Press, 1974), pp. 43–44.

39. Hubert Schmidt, *Policy and Functioning in Industry* (Karlsruhe, West Germany: U.S. Military Government in Germany, 1950), p. 13.

40. Clayton to Secretary of Labor, October 5, 1945, 840.6362/10-5-45, RG 59.

41. Stephen D. Krasner, "United States Commercial Policy: Unraveling the Paradox of External Strength and Internal Weakness," in Peter Katzenstein, ed., *Between Power and Plenty* (Madison: University of Wisconsin Press, 1978), pp. 51–88.

42. Ickes to Clayton, October 18, 1945, 840.6362/10-18-45; Clayton to Ickes, October 25, 1945, 840.6362/10-25-45, RG 59. The primary purpose of Ickes' trip

was to negotiate an Anglo-American oil agreement with the British. See Painter, *Oil and the American Century,* pp. 69–70.

43. London Embassy to Secretary of State, December 22, 1945, 840.6362/10-22-45, RG 59.

44. Warsaw Embassy to Secretary of State, February 10, 1945, 840.6362/2-10-46, RG 59. For a theoretical discussion of "norm-governed change" in international organizations, see John Ruggie, "International Regimes, Transactions, and Change," in Krasner, ed., *International Regimes*, pp. 195–231.

45. Paul Porter to the Secretary of State, January 31, 1949, 501.BD Europe-Coal, RG 59. On Polish–U.S. relations, see P.S. Wandycz, *The United States and Poland* (Cambridge, Mass.: Harvard University Press, 1980).

46. F. A. Lightner, "Russian Interest in the Status of the Ruhr and Effect on U.S. Policy," January 21, 1946, Records of the Central European Division, RG 59, box 2.

47. Ibid.

48. Galbraith to Clayton, April 30, 1946, 840.6362/4-30-46, RG 59.

49. Acheson and Clayton to Byrnes, May 15, 1946, 840.6362/5-15-46, RG 59.

50. Alan Milward, *The Reconstruction of Western Europe* (London: Methuen, 1984), p. 129.

51. Monnet to Clayton, May 18, 1946, 840.6362/5-18-46, RG 59.

52. Pollard, *Economic Security*, p. 57.

53. Kindleberger, "Toward the Marshall Plan," p. 14.

54. ECO, *European Coal Organization*.

55. Harriman to Secretary of State, June 13, 1946, 840.6362/6-13-46, RG 59.

56. Byrnes to Harriman, July 9, 1946, 840.6362/7-9-46, RG 59.

57. Louis Lister to K. Anderson, October 18, 1946, 840.6362/10-18-46, RG 59.

58. Gridley to Lie, November 7, 1946, 840.6362/11-7-46, RG 59.

59. Walter LaFeber, *America, Russia, and the Cold War* (New York: Wiley, 1972), p. 43; Conservative Party Central Office, *Fuel and Power Crisis* (London, February 24, 1947); ECO, *European Coal Organization*.

60. See Pollard, *Economic Security*, pp. 107–132; he notes that Turkey was perceived by U.S. officials as "strategic" in part because of its proximity to the Iranian oil fields.

61. Scott Jackson, "Prologue to the Marshall Plan," *Journal of American History* (March 1979): 1043–1068.

62. Policy Planning Staff, Paper 1, May 23, 1947, Records of the Policy Planning Staff, RG 59.

63. Policy Planning Staff, Paper 2, June 2, 1947, RG 59.

64. On Acheson's Delta Speech, see Pollard, *Economic Security,* p. 135. Will Clayton, "The European Crisis," May 31, 1947, Will Clayton Papers, box 60, HSTL.

65. The speech is in *FRUS: 1947*, v. 3, pp. 237–239.

66. For a recent study of the plan see Michael Hogan, *The Marshall Plan* (New York: Cambridge University Press, 1987).

67. Pollard, *Economic Security*, p. 139.

68. *New York Times*, September 11, 1947, p. 1.

69. See, e.g., "Coal in Europe," *The World Today 4* (March 1948): 111-117.

Chapter 3

1. Quoted in Walter Millis, ed., *The Forrestal Diaries* (New York: Viking, 1951), p. 272.

2. Hadley Arkes, *Bureaucracy, the Marshall Plan and the National Interest* (Princeton, N.J.: Princeton University Press, 1972); Ethan B. Kapstein, "The Marshall Plan and Industrial Policy," *Challenge* (May/June 1984): 55–59.

3. David Painter, "Oil and the Marshall Plan," *Business History Review 58* (Autumn 1984): 360; See also Melvyn P. Leffler, "The United States and the Strategic Dimensions of the Marshall Plan," *Diplomatic History 12* (Summer 1988): 277–306.

4. Louis Lister, "Recovery of Coal Production in Europe," July 10, 1947, Records of Interdepartmental Committees, RG 353, box 27.

5. Louis Lister to S. W. Anderson, August 24, 1948, Records of the Agency for International Development, RG 286, box 160.

6. U.S. Economic Cooperation Administration (ECA), *Coal and Related Solid Fuels Commodity Study*, (Washington: Government Printing Office, 1949).

7. Ibid.

8. Paul Porter to Will Clayton, November 19, 1946, 860c.6362/11-1946, RG 59.

9. Lister, "Recovery of Coal Production."

10. Charles E. Bohlen, "Summary of Department's Position on the Content of a European Recovery Program," August 26, 1947, Records of Charles E. Bohlen, RG 59, box 6.

11. Paul Porter to Secretary of State, January 31, 1949, 501.BD Europe-Coal/1-3149, RG 59.

12. See Geir Lundestad, *The American Non-Policy Towards Eastern Europe, 1943–1947* (New York: Humanities Press, 1975).

13. ECO, "Draft Document on Provisional Estimate of Dollar Requirements," July 18, 1947, RG 353, box 27.

14. Charles Kindleberger, "Role of Coal in U.S. Aid to a European Recovery Program," July 29, 1947, RG 353, box 27.

15. Committee for European Economic Cooperation (CEEC), *Technical Reports* (Washington: Government Printing Office, 1947), pp. 132–180.

16. Washington Conversations on European Economic Cooperation (hereafter: WC), "Interim Report on Mining Machinery Conversations," October 13, 1947, RG 353, box 27.

17. Ibid.

18. Bohlen, "Summary of Department's Position"; Armin Rappoport, "The United States and European Integration: The First Phase," *Diplomatic History 5* (Spring 1981): 121–149.

19. WC, "Summary Record of Electric Power Conversations," October 7, 1947, RG 353, box 27.

20. H. W. Howell, memorandum, May 12, 1948, Records of the Bipartite Control Organization, RG 260, entry BCO, file ERP. For a secondary account that takes this approach, see Gimbel, *Origins of the Marshall Plan*.

21. U.S. High Commissioner for Germany, "The West German Coal Industry

Since 1945," November 14, 1951, Records of International Conferences, RG 43, box 288.

22. John M. Rogers, *The International Authority for the Ruhr* (Ph.D. dissertation, American University, 1960), p. 189.

23. Anton DePorte, *Europe Between the Superpowers* (New Haven, Conn.: Yale University Press, 1979), p. 222; see also Raymond Vernon, "The Schuman Plan," January 5, 1951, RG 43, box 298; at this time Vernon was a State Department economist serving as Deputy Director, Office of Economic Defense and Trade Policy.

24. William Diebold, *The Schuman Plan* (New York: Praeger, 1959), p. 324.

25. Rappoport, "The United States and European Integration."

26. Diebold, *The Schuman Plan*, p. 558.

27. U.S. High Commissioner for Germany, "The Schuman Plan," 1951, RG 43, box 298.

28. That the ECSC had "learned" from the ECO was noted by Nathaniel Samuels in an interview with the author, April 8, 1985.

29. Ibid.; Samuels helped to sell the first ECSC bond issues.

30. Freymond, *The Saar Conflict*, p. 40.

31. Ibid., p. 303.

32. William Haynes, *Nationalization in Practice: The British Coal Industry* (Boston: Harvard Business School Press, 1953); A. R. Griffin, *The British Coalmining Industry* (Buxton, U.K.: Moorland, 1977).

33. Organization for European Economic Cooperation (OEEC), *The Coal Industry in Europe* (Paris: OEEC, 1954).

34. Walter Millis, ed., *The Forrestal Diaries* (New York: Viking, 1951), pp. 356–357.

35. Donald Calder, "Progress in British Conversion to Petroleum Fuel," February 14, 1947, Records of Foreign Service Posts, RG 84, file 863.6, box 1023.

36. Dalton to George Marshall, April 16, 1947, RG 84, file 863.6, box 1023.

37. Clayton to Sawyer, July 31, 1946, 855.6362/7-31-46.

38. U.S. Congress. House Committee on Interstate and Foreign Commerce, *Fuel Investigation: Petroleum and the European Recovery Program*, H.R. 1438, 80th Cong., 2d sess., 1948.

39. Willard Thorp to George Kennan, July 9, 1947, RG 353, box 27.

40. Petroleum Division, "Foreign Petroleum Policy of the United States," April 11, 1944, Petroleum Division files, RG 59, box 1; also in *FRUS* 1944, v. 5, pp. 27–33.

41. Quoted in Craufurd Goodwin, *Energy Policy in Perspective* (Washington: Brookings Institution, 1981) p. 77.

42. Stephen Longrigg, *Oil in the Middle East* (New York: Oxford University Press, 1954), p. 277.

43. Charles Issawi and Mohammed Yeganeh, *The Economics of Middle East Oil* (New York: Frederick Praeger, 1962); Edith Penrose, *The International Oil Industry in the Middle East* (Cairo: National Bank of Egypt, 1968).

44. "Up from the Ashes," *The Lamp* (September 1951).

45. Walter LaFeber, *America, Russia, and the Cold War* (New York: John Wiley and Sons, 1972), p. 28.

46. CEEC, *General Report* (Washington: U.S. State Department, 1947), p. 90.

47. Washington Conversations, "Interim Report on Petroleum Conversations," October 13, 1947, RG 353, box 27.

48. Ibid.

49. Arkes, *Bureaucracy*; William Adams Brown Jr. and Redvers Opie, *American Foreign Assistance* (Washington: Brookings Institution, 1953).

50. Lincoln Gordon, "The ERP in Operation," *Harvard Business Review 27* (March 1949): 129–150.

51. ECA, "Effects of Imports of Foreign Oil on the Domestic Independent Producers," December 9, 1949, RG 286, entry 531405, box 33.

52. U.S. Congress, *Fuel Investigation*.

53. Arkes, *Bureaucracy*, p. 134.

54. Quoted in Aaron David Miller, *Search for Security* (Chapel Hill, N.C.: University of North Carolina Press, 1980), p. 197.

55. Millis, ed., *The Forrestal Diaries*, pp. 323–324.

56. Quoted in Irvine Anderson, *Aramco, the United States, and Saudi Arabia* (Princeton, N.J.: Princeton University Press, 1981), p. 177.

57. U.S. Congress, *Fuel Investigation*.

58. On changing market conditions see ECA, "Effects of Imports" and Painter, *Oil and the American Century*, pp. 99–100.

59. Walter Levy, memo, August 13, 1949, RG 286, box 153, for the connection between British financial and oil policies.

60. Leo Welch to Paul Nitze, April 22, 1949, 840.6363/4-22-49, RG 59.

61. Joint Petroleum Discussion Papers, 1949, RG 353, box 83.

62. George Walden to Huntington Gilchrist, December 1949, RG 286, box 153.

63. Paul Hoffman to OSR, December 19, 1949, RG 286, box 153.

64. See Michael Stoff, *Oil, War, and American Security* (New Haven, Conn.: Yale University Press, 1980).

65. Hollis Chenery to George Walden, February 21, 1950, RG 286, box 150; George Walden to Huntington Gilchrist, February 23, 1950, RG 286, box 153.

66. Brookings Institution, *Current Developments in World Affairs 3* (February 1950): 11.

67. Robert Keohane, "Hegemonic Leadership and U.S. Foreign Economic Policy in the 'Long Decade' of the 1950s," in William Avery and David Rapkin, eds., *America in a Changing World Political Economy* (New York: Longmans, 1982), pp. 59–60.

68. Willard Thorp, "Aspects of International Petroleum Policy," *Department of State Bulletin*, April 25, 1950, pp. 640–645.

69. "Oil's Role in Britain's Struggle," *The Lamp* (September 1953): 2–5.

70. Horst Mendershausen, "Dollar Shortage and Oil Surplus in 1949–50," *Essays in International Finance 11* (November 1959).

71. ECA, "ECA's Petroleum Policy," June 26, 1950, RG 286, box 33.

72. E. Groen, "The Significance of the Marshall Plan for the Petroleum Industry in Europe," *Report of the Third World Petroleum Congress* (The Hague, 1951), pp. 38–66; Leonard Fanning, *Foreign Oil and the Free World* (New York: McGraw-Hill, 1954), p. 345.

73. Richard Funkhouser, "Middle East Oil," September 1950, *FRUS: 1950,* v. V, p. 82.

Chapter 4

1. Dean Acheson, *Present at the Creation* (New York: Norton, 1969), p. 505.

2. Quoted in George C. McGhee, *Envoy to the Middle World* (New York: Harper & Row, 1983), p. 398.

3. Richard Funkhouser, "Middle East Oil," September 1950, *FRUS: 1950,* v. V, p. 82.

4. Ibid.

5. Ibid.

6. Richard Funkhouser, "Discussions with British on AIOC," September 14, 1950, *FRUS: 1950*, v. V, p. 98.

7. Quoted in Ronnie Lipschutz, *Ore Wars: Access to Strategic Minerals, International Conflict, and the Foreign Policies of States* (Ph.D. dissertation, University of California, Berkeley, 1987), p. 318.

8. National Security Council, "Position of the United States with Respect to Iran," June 20, 1951, NSC 107/1, *Declassified Documents Reference System* (DDRS), 1983, doc. 1293.

9. Philip Jessup to Secretary of State, August 26, 1950, in *FRUS: 1950*, v. V, p. 611. For a secondary account that emphasizes the importance of British military power, see William Stivers, *America's Confrontation with Revolutionary Change in the Middle East, 1945–83* (London: Macmillan Press, 1986).

10. Council on Foreign Relations, *The United States in World Affairs: 1951* (New York: Harper & Brothers, 1952), p. 267.

11. Peter Calvocoressi, *Survey of International Affairs: 1951* (London: Oxford University Press, 1954), p. 305.

12. McGhee, *Envoy to the Middle World*, p. 335.

13. Ibid., p. 306.

14. Dean Acheson, *Present at the Creation,* p. 506.

15. Richard Cottam, *Nationalism in Iran* (Pittsburgh, Pa.: University of Pittsburgh Press, 1964), p. 206; see also James A. Bill, *The Eagle and the Lion* (New Haven, Conn.: Yale University Press, 1988), ch. 2.

16. Calvocoressi, *Survey: 1951*, p. 309.

17. Council on Foreign Relations, *United States in World Affairs: 1951*, p. 273.

18. Acheson, *Present*, pp. 507–508.

19. Acheson, *Present*, pp. 508–509; Calvocoressi, *Survey: 1951*, p. 321.

20. Ibid.

21. Funkhouser, "Discussions with British."

22. Painter, *Oil and the American Century*, pp. 180–181.

23. OEEC Oil Committee, *Third Report on Coordination of Oil Refinery Expansion in the OEEC Countries* (Paris: OEEC, April 1953), p. 16.

24. Organisation for European Economic Cooperation (OEEC) Oil Committee, Internal Files, 1951, International Energy Agency (IEA) Archives, Paris.

25. United Nations, Economic Commission for Europe, "The Price of Oil in Western Europe," March 14, 1955.

26. Quoted in Michael Tanzer, *The Political Economy of Oil and the Underdeveloped Countries* (Boston: Beacon Press, 1969), p. 322.

27. David Painter, "The United States and Iran, 1951–1953: The Political Econ-

omy of Intervention," paper presented at the American Historical Association Conference, Honolulu, August 15, 1986.

28. OEEC Oil Committee, Internal Files, 1952, IEA.

29. OEEC, *Developpement du Raffinage du Petrole* (Paris: OEEC, 1949).

30. OEEC Oil Committee, Internal Files, 1953, IEA.

31. Federal Trade Commission, *The International Petroleum Cartel* (Washington: Government Printing Office, 1952).

32. See Burton I. Kaufman, *The Oil Cartel Case* (Westport, Conn.: Greenwood Press, 1978).

33. Ibid.

34. Painter, "The United States and Iran," p. 17.

35. Acheson, *Present*, p. 679.

36. Peter Calvocoressi, *Survey of International Affairs: 1953* (London: Oxford University Press, 1956), pp. 249–259.

37. Acheson, *Present*, p. 681.

38. Ibid., p. 682.

39. Ibid.

40. Ibid., p. 684.

41. Ibid.; Painter, "The United States and Iran," p. 23.

42. Quoted in Barry Rubin, *Paved with Good Intentions* (New York: Oxford University Press, 1980), p. 77.

43. Kermit Roosevelt, *Countercoup* (New York: McGraw Hill, 1979). Secondary accounts are found in numerous studies including Painter, "The United States and Iran," and *Oil and the American Century*; Rubin, *Paved*; and Bill, *The Eagle and the Lion*.

44. Richard Saunders, "Military Force in the Foreign Policy of the Eisenhower Presidency," *Political Science Quarterly 100* (Spring 1985), p. 115.

45. Council on Foreign Relations, *The United States in World Affairs: 1953* (New York: Harper & Brothers, 1955), p. 306.

46. Ibid.

47. Painter, "The United States and Iran," pp. 26–28. Painter argues that U.S. interest in a coup d'état rose as the international oil boycott began to weaken in the spring of 1953.

48. Rubin, *Paved*, p. 82.

49. Painter, "The United States and Iran," p. 29.

50. Council on Foreign Relations, *United States in World Affairs: 1953*, p. 308.

51. U.S. Congress, Senate Committee on Foreign Relations, *Multinational Oil Corporations and U.S. Foreign Policy* (Washington: Government Printing Office, 1975), pp. 64–74.

52. Ibid., p. 65.

53. Richard Funkhouser, "Middle East Oil Policy Considerations," July 3, 1953 (rev. September 10, 1953), in U.S. Congress, Senate Committee on Foreign Relations, *Multinational Corporations and U.S. Foreign Policy: Part 7*, 93d Congress, 2d session, (Washington: Government Printing Office, 1974).

54. Committee on Foreign Relations, *Multinational Oil*, p. 66.

55. For a detailed account of the consortium see Stephen Longrigg, *Oil in the Middle East* (London: Oxford University Press, 1968), pp. 277–289.

56. Ibid., pp. 282–284.

57. Council on Foreign Relations, *United States in World Affairs: 1953*, p. 309.

Chapter 5

1. Quoted in Department of State, *The Suez Canal Problem* (Washington: Government Printing Office, 1956), p. 76.

2. Council on Foreign Relations, *United States in World Affairs: 1952*, p. 282.

3. Burton Berry to Dean Acheson, "Problems with Regard to Egypt," October 12, 1950, *FRUS 1950*, v. V, p. 305.

4. Wells Stabler, "Memorandum of Conversation," July 17, 1950, *FRUS: 1950*, v. V, p. 294. See also Stivers, *American Confrontation,* pp. 10–11.

5. Stuart Nelson to Burton Berry, "Problem Summary for Egypt," February 24, 1950, *FRUS: 1950*, v. V, p. 284.

6. Wells Stabler to Burton Berry, "Anglo-Egyptian Negotiations," June 1, 1950, *FRUS: 1950*, v. V, p. 291.

7. Council on Foreign Relations, *United States in World Affairs: 1952*, p. 237.

8. Ibid.

9. See the "Biographical Note" by John Gunther in Gamal Abdel Nasser, *The Philosophy of the Revolution* (Buffalo, N.Y.: Economica Books, 1959).

10. Herman Finer, *Dulles Over Suez* (Chicago: Quadrangle Books, 1964), p. 16.

11. "Convention Respecting the Free Navigation of the Suez Maritime Canal," Constantinople, October 29, 1888, in Department of State, *The Suez Canal Problem*, p. 17.

12. Stuart Nelson to Burton Berry, "Problem Summary for Egypt," February 24, 1950, *FRUS: 1950*, v. V, p. 286.

13. Geoffrey Barraclough and Rachel Wall, *Survey of International Affairs: 1955–1956* (London: Oxford University Press, 1960), p. 96.

14. On the dam see Ibid.; Finer, *Dulles*, pp. 37–54; and Hugh Thomas, *Suez* (New York: Harper & Row, 1967), pp. 22–24.

15. Ibid.

16. Finer, *Dulles*, p. 54. Finer argues that Dulles acted without Eisenhower's prior knowledge. This assertion is disputed by a former intelligence officer deeply involved in the Suez dispute, Chester L. Cooper. See Cooper, *The Lion's Last Roar: Suez 1956* (New York: Harper & Row, 1978), p. 98.

17. On the history see Cooper, *Lion's Last Roar*.

18. OEEC Oil Committee, Internal Files, 1954, IEA.

19. Quoted in Thomas, *Suez*, p. 33.

20. Quoted in Finer, *Dulles*, p. 63.

21. Ibid.; see also Thomas, *Suez*, p. 33; Cooper, *Lion's Last Roar*, pp. 109–110.

22. Geoffrey Barraclough, *Survey of International Affairs: 1956–1958* (London: Oxford University Press, 1962), p. 8.

23. Thomas, *Suez*, p. 19.

24. Barraclough, *Survey*, p. 17.

25. Cooper, *Lion's Last Roar*, p. 112. For a study that focuses on Eisenhower's legal concerns see Robert R. Bowie, *Suez 1956* (New York: Oxford University Press, 1974). Bowie was director of the Policy Planning Staff at the State Department and a close adviser to Dulles.

26. Cooper, *Lion's Last Roar*, p. 124.

27. Arthur Flemming to Fred Seaton, July 31, 1956, in U.S. Congress, Senate, *Joint Hearings Before Subcommittees of the Committee on the Judiciary and the Committee on Interior and Insular Affairs: Emergency Oil Lift Program and Related Oil Problems* (Washington: Government Printing Office, 1957), p. 574.

28. Seaton to Flemming, August 1, 1956, in Ibid.

29. "Plan of Action Under Voluntary Agreement Relating to Foreign Petroleum Supply," August 10, 1956, in Ibid., pp. 578–582.

30. OEEC, *Europe's Need for Oil* (Paris: OEEC, 1958), p. 20.

31. OEEC Oil Committee, Internal Files, 1956, IEA.

32. "Bipartisan Leadership Meeting," August 12, 1956, *DDRS* (1976), doc. 217B.

33. Ibid.

34. Ibid.

35. The conferences speeches are in Department of State, *The Suez Canal Problem.*

36. Ibid., pp. 76–77.

37. The correspondence is in Ibid.

38. Quoted in Cooper, *Lion's Last Roar*, p. 126.

39. Ibid.

40. Anthony Eden formally proposed the SCUA idea in a speech before the House of Commons on September 12, and in his news conference of September 13 Dulles said it was Eden's "concept." But it was widely known at the time that it was really Dulles's creation.

41. The texts on SCUA are in Department of State, *The Suez Canal Problem.*

42. Barraclough, *Survey: 1956–1958*, p. 39.

43. Finer, *Dulles*, p. 216.

44. Barraclough, *Survey: 1956–1958*, p. 39.

45. Cooper, *Lion's Last Roar*, p. 126.

46. Finer, *Dulles*, p. 328.

47. Ibid., pp. 328–332.

48. Barraclough, *Survey: 1956–1958*, p. 57.

49. Ibid.

50. Ibid., p. 62.

51. Richard K. Betts, *Nuclear Blackmail and Nuclear Balance* (Washington: Brookings Institution, 1987), pp. 62–63.

52. Cooper, *Lion's Last Roar*, p. 197.

53. Finer, *Dulles*, p. 411.

54. Quoted in Betts, *Nuclear Blackmail*, p. 64.

55. Finer, *Dulles*, p. 122.

56. Quoted in William J. Barber, "The Eisenhower Energy Policy: Reluctant Intervention," in Craufurd Goodwin, ed., *Energy Policy in Perspective* (Washington: Brookings Institution, 1981), p. 235.

57. Anthony Adamthwaite, "Suez Revisited," *International Affairs 64* (Summer 1988), p. 455.

58. Barraclough, *Survey: 1956–1958*, p. 69.

59. Quoted in Finer, *Dulles*, p. 446.

60. OEEC, *Europe's Need for Oil*, p. 23.

61. Ibid.

62. Ibid., p. 34.

63. U.S. Congress, Senate, *Emergency Oil Lift*, p. 104; Douglas Bohi and Milton Russell, *Limiting Oil Imports* (Baltimore: Johns Hopkins University Press, 1978), pp. 37–40.

64. OEEC, *Europe's Need for Oil*, p. 30.

65. Melvin de Chazeau and Alfred Kahn, *Integration and Competition in the Petroleum Industry* (New Haven, Conn.: Yale University Press, 1959), p. 198.

66. Krasner, "United States Commercial Policy."

67. OEEC Oil Committee, Internal Files, 1957, IEA.

68. OEEC, *Europe's Need for Oil*, p. 34.

69. OEEC, *Europe's Need for Oil*, pp. 39–42.

70. OEEC Oil Committee, Internal Files, 1957, OEEC Archives, Paris.

71. OEEC, *Europe's Need for Oil*, p. 44.

72. Walter Levy, "Issues in International Oil Policy," *Foreign Affairs* (April 1957), pp. 10–11.

73. Ibid.

74. Ibid., p. 13.

75. Sauders, "Military Force in Eisenhower's Foreign Policy," p. 111.

76. Campbell, *Defense of the Middle East*, p. 110.

77. Gaddis, *Strategies of Containment*, p. 180; Cottam, *Nationalization in Iran*, p. 233.

78. Stephen E. Ambrose, *Rise to Globalism* (New York: Penguin Books, 1983), p. 217.

Chapter 6

1. Quoted in Christopher Rand, *Making Democracy Safe for Oil* (Boston: Little, Brown, 1975), p. 98.

2. Harold Lubell, "Security of Supply and Energy Policy in Western Europe," *World Politics 13* (April 1961), p. 402.

3. Arnold Kramish, *The Peaceful Atom in Foreign Policy* (New York: Harper & Row, 1963), p. 39.

4. See Lawrence Scheinman, *Atomic Energy Policy in France Under the Fourth Republic* (Princeton, N.J.: Princeton University Press, 1965), pp. 129–165.

5. National Security Council, *Peaceful Uses of Atomic Energy*, NSC 5507/2, March 12, 1955, Department of Energy (DOE) Archives, Germantown, Md., 1951–1958 Secretariat Files, R & D 1, International Controls.

6. Dulles to Strauss, January 16, 1956, DOE Archives, 1951–1958 Secretariat Files, R & D 1, Peacetime Uses.

7. Atomic Energy Commission, *European Integration in the Field of Atomic Energy*, AEC 751/55, March 14, 1956, DOE, 1951–1958 Secretariat Files, R & D 1, Peacetime Uses.

8. Kramish, *Peaceful Atom*, p. 153.

9. Gerard Smith, "Report on the Status of Euratom," January 7, 1957, AEC 751/102, DOE, 1951–1958 Secretariat Files, R & D 1, Peacetime Uses.

10. Kramish, *Peaceful Atom*, p. 88.

11. Smith, *Euratom*.

12. John Foster Dulles to Lewis Strauss, February 5, 1957, AEC 751/110, DOE, 1951–1958 Secretariat Files, R & D 1, Peacetime Uses.

13. Kramish, *Peaceful Atom*, p. 157.

14. Scheinman, *Atomic Energy Policy*, p. 171; see also Warren H. Donnelly, *Commercial Nuclear Power in Europe: The Interaction of American Diplomacy with a New Technology* (Washington: Government Printing Office, 1972).

15. Quoted in Lubell, "Security of Supply," p. 417.

16. OEEC Oil Committee, Internal Files, 1958, IEA.

17. OECD Oil Committee, Internal Files, 1961, IEA. As will be discussed later, in 1961 the OEEC was transformed into the OECD.

18. OEEC Oil Committee, Internal Files, 1958, IEA.

19. Ibid.

20. OEEC Oil Committee, Internal Files, 1958, IEA.

21. OECD Oil Committee, Internal Files, 1964, IEA.

22. Lubell, "Security of Supply," pp. 414–417.

23. Frank, *Crude Oil Prices in the Middle East*, p. 95.

24. Joel Darmstadter and Hans Ladsberg, "The Economic Background," in Raymond Vernon, ed., *The Oil Crisis* (New York: Norton, 1976), p. 22.

25. Richard Vietor, *Energy Policy in America* (New York: Cambridge University Press, 1984), pp. 91-115.

26. Ibid., p. 108.

27. U.S. Cabinet Task Force on Oil Import Control, *The Oil Import Question* (Washington: Government Printing Office, February 1970), p. 11.

28. Bohi and Russell, *Limiting Oil Imports*, p. 64.

29. Frank Wyant, "The Role of Multinational Oil Companies in World Energy Trade," in Jack Hollander, ed., *Annual Review of Energy* (Palo Alto, Calif.: Annual Reviews, 1977), pp. 136–137.

30. Ibid.; U.S. Cabinet, *Oil Import Question*, p. 6.

31. Robert O. Keohane, "Hegemonic Leadership and U.S. Foreign Economic Policy in the 'Long Decade' of the 1950s," in William P. Avery and David P. Rapkin, *America in a Changing World Political Economy* (New York: Longman, 1982), pp. 70–71.

32. On this rationale see National Security Council, "Means of Retarding Western Europe's Increasing Reliance on Middle East Oil," June 23, 1960, *DDRS* (1981), doc. 338A.

33. See Bohi and Russell, *Limiting Oil Imports*, for an economic comparison of the alternatives.

34. Quoted in NSC, "Western Europe's Reliance on Middle East Oil."

35. Ibid.

36. Department of State, Bureau of Intelligence and Research, *Economic and Political Significance of North African Oil Discoveries*, Intelligence Report No. 8091, August 27, 1959, *DDRS* (1979), doc. 61B.

37. Ibid.

38. Ibid.; W. G. Jensen, *Energy in Europe: 1945–1980* (London: G. T. Foulis, 1967), pp. 58–64.

39. NSC, "Western Europe's Reliance on Middle East Oil."

40. Ibid.

41. J. E. Hartshorn, *Politics and World Oil Economics* (New York: Praeger, 1962), p. 217.

42. Peter Odell, *Oil and World Power* (Baltimore: Penguin Books, 1974), p. 52.

43. Frank, *Crude Oil Prices in the Middle East,* pp.103–113.

44. John Scott, *The Soviet Economic Offensive* (New York: Time Books, 1961), pp. 114–119.

45. Odell, *Oil and World Power*, p. 55.

46. On Soviet–European oil agreements after Suez see Bruce Jentleson, *Pipeline Politics* (Ithaca, N.Y.: Cornell University Press, 1986), pp. 76–131.

47. Bruce Jentleson, "From Consensus to Conflict: The Domestic Political Economy of East–West Energy Trade," *International Organization 38* (Autumn 1984), p. 637.

48. Ibid.

49. Dankwart Rustow and John Mugno, *OPEC: Success and Prospects* (New York: New York University Press, 1976), p. 5.

50. William Diebold, "Economic Aspects of an Atlantic Community," in Francis O. Wilcox and H. Field Haviland, eds., *The Atlantic Community: Progress and Prospects* (New York: Praeger, 1963), p. 156.

51. Ibid.

52. Henry G. Aubrey, *Atlantic Economic Cooperation* (New York: Praeger, 1967), pp. 29–30.

53. Ibid., p. 78.

54. OECD Oil Committee, *Report on the Stockpiling Programme* (Paris: OECD, 1961).

55. OECD Oil Committee, "Draft Preliminary Report on the 1967 Oil Emergency," August 28, 1968, cited in Fred Tanner, *Energy and Alliance Tensions* (Ph.D. dissertation, Fletcher School of Law and Diplomacy, 1984).

56. Ibid.

57. Richard P. Stebbins, *The United States in World Affairs: 1967* (New York: Simon & Schuster, 1968), p. 86.

58. Fred J. Khouri, *The Arab–Israeli Dilemma* (Syracuse, N.Y.: Syracuse University Press, 1968), p. 245.

59. Quoted in Stebbins, *United States in World Affairs: 1967*, p. 93.

60. White House, "United States Policy and Diplomacy in the Middle East Crisis, May 15–June 10, 1967," *DDRS* (1984), doc. 2873.

61. Sylvia K. Crosbie, *A Tacit Alliance* (Princeton, N.J.: Princeton University Press, 1974), p. 191.

62. Stebbins, *United States in World Affairs: 1967*, p. 98.

63. For details on the Six Day War see Nadav Safran, *From War to War* (New York: Pegasus, 1969).

64. American Embassy (London) to Dean Rusk, June 6, 1967, *DDRS* (1984), doc. 602; American Embassy (Tripoli) to Dean Rusk, June 8, 1967, *DDRS* (1984), doc. 604; OECD Oil Committee, "Draft Preliminary Report."

65. White House, "United States Policy."

66. Ibid.

67. Central Intelligence Agency, "Impact on Western Europe and Japan of a Denial of Arab Oil," June 7, 1967, *DDRS* (1981), doc. 280A.

68. Anthony M. Solomon, "The Middle Eastern Oil Problem," June 9, 1967, *DDRS* (1981), doc. 569C.

69. Ibid.

70. OECD Oil Committee, "Draft Preliminary Report;" "Coping with the Crisis," *Petroleum Press Service* (July 1967), pp. 244–245.

71. OECD Oil Committee, Internal Files, 1967, IEA.

72. Quoted in Walter Laquer, *The Road to War* (Harmondsworth, U.K.: Penguin Books, 1968), p. 195.

73. Stebbins, *United States in World Affairs: 1967*, p. 202.

74. Ibid.

75. 'Feat of Redeployment," *Petroleum Press Service* (August 1967), pp. 282–283.

76. "Nigerian Oil in the Balance," *Petroleum Press Service* (June 1967), pp. 208–210; "Civil War Halts Output," *Petroleum Press Service* (August 1967), p. 306.

77. "Feat of Redeployment," *Petroleum Press Service* (August 1967), pp. 282–283; U.S. Department of the Interior, Emergency Petroleum Supply Committee, "Appraisal of Foreign Petroleum Supply/Demand—3rd Quarter 1967," August 15, 1967.

78. Stebbins, *United States in World Affairs: 1967*, p. 122.

79. OECD Oil Committee, "Draft Preliminary Report."

80. See "Whither OPEC?," *Petroleum Press Service* (October 1967), pp. 362–363.

Chapter 7

1. Walter Laquer, *Confrontation* (New York: Quadrangle Books, 1972), p. 223.

2. For extended discussions of oil markets in the early 1970s see Steven A. Schneider, *The Oil Price Revolution* (Baltimore: Johns Hopkins University Press, 1983); Raymond Vernon, ed., *The Oil Crisis* (New York: Norton, 1976); Louis Turner, *Oil Companies in the International System* (London: Allen & Unwin, 1978).

3. Department of State, Office of the Historian, "OPEC and the Changing Structure of the World Oil Economy, 1960–1983," *Historical Issues 10* (1973).

4. Mira Wilkins, "The Oil Companies in Perspective," in Vernon, ed., *The Oil Crisis*, pp. 167–168.

5. Ibid.

6. State Department, "OPEC."

7. OECD Oil Committee, Internal Files, 1970, IEA; also see Joel Darmstadter and Hans Landsberg, "The Economic Background," in Vernon, ed., *The Oil Crisis*.

8. Joseph Kalt, *The Economics and Politics of Oil Price Regulation* (Cambridge, Mass.: MIT Press, 1981), p. 8.

9. U.S. Cabinet Task Force on Oil Import Control, *The Oil Import Question* (Washington: Government Printing Office, 1970), p. 131.

10. Ibid., p. 32.

11. Quoted in Congressional Quarterly, *Energy Crisis in America* (Washington: *Congressional Quarterly,* 1973), p. 84.

12. Kalt, *Economics and Politics*, p. 289.

13. Goodwin, *Energy Policy in Perspective*, p. 693.

14. OECD Oil Committee, Internal Files, 1969, IEA.

15. OECD Oil Committee, Internal Files, 1970, IEA.

16. OECD Oil Committee, *Report on Oil Stockpiling* (Paris: OECD, 1971).

17. Quoted in Walter Levy, "An Atlantic–Japanese Energy Policy," in Richard Mayne, ed., *The New Atlantic Challenge* (New York: John Wiley & Sons, 1975), pp. 142–143.

18. OECD Oil Committee, Internal Files, 1972, IEA.

19. Schneider, *Oil Price Revolution*, p. 193.

20. Valerie Yorke, "Oil, the Middle East, and Japan's Search for Security," in Nobitushi Akao, ed., *Japan's Economic Security* (New York: St. Martin's Press, 1983), p. 53.

21. Mark Brown, "The Emergence of Japanese Interests in the World Oil Market," Occasional Paper 83-01, U.S.–Japan Program, Harvard Center for International Affairs, 1983, p. 65.

22. OECD Oil Committee, *Oil Demand and Supply: Problems and Prospects to 1980* (Paris: OECD, 1973).

23. Shell Group Planning, "Measures to Mitigate an Oil Scramble and Methods of Allocating Oil Supplies," May 1973, mimeograph.

24. Robert Ebel, "Alternative Criteria for Sharing Oil Supplies," U.S. Department of Interior, July 1973, mimeograph.

25. Ulf Lantzke, "The OECD and its International Energy Agency," in Vernon, ed., *The Oil Crisis*, p. 218.

26. Joan Garratt, "Euro–American Energy Diplomacy in the Middle East, 1970–80: The Pervasive Crisis," in Steven L. Spiegel, ed., *The Middle East and the Western Alliance* (London: George Allen & Unwin, 1982), pp. 87–97.

27. Marshall Goldman, "The Soviet Union," in Vernon, *The Oil Crisis*, pp. 133–137.

28. Charles A. Kupchan, *The Persian Gulf and the West* (Boston: Allen & Unwin, 1987).

29. Khouri, *The Arab–Israeli Dilemma*, pp. 319–355. See also Raymond L. Garthoff, *Detente and Confrontation* (Washington: Brookings Institution, 1985).

30. The chronology of events is detailed in Congressional Quarterly, *The Middle East* (Washington: Congressional Quarterly, 1977).

31. Stephen M. Walt, *The Origins of Alliances* (Ithaca, N.Y.: Cornell University Press, 1987), pp. 114–128.

32. See Walter Laquer, *Confrontation* (New York: Quadrangle Books, 1974).

33. Schneider, *Oil Price Revolution*, p. 217; Yusif Sayigh, "Oil in Arab Developmental and Political Strategy: An Arab View," in John Duke Anthony, ed., *The Middle East: Oil, Politics, and Development* (Washington: American Enterprise Institute, 1975).

34. Laquer, *Confrontation*, p. 55.

35. See Laquer, *Confrontation*; and Nadav Safran, "The War and the Future of the Arab–Israeli Conflict," *Foreign Affairs* (January 1974). For a good summary treatment, see Robert J. Lieber, "American Diplomatic Response to the 1973–74 Energy Crisis," Washington: Georgetown University, August 20, 1988, mimeograph.

36. Federal Energy Administration, *U.S. Oil Companies and the Arab Oil Embargo* (Washington, D.C.: Government Printing Office, 1975).

37. Laquer, *Confrontation*, pp. 168–169.

38. Sayigh, "Oil," p. 43.

39. Quoted in FEA, *U.S. Oil Companies*.

40. Laquer, *Confrontation*, pp. 175–181.

41. FEA, *U.S. Oil Companies*.

42. Garratt, "Euro–American Energy Diplomacy," p. 84.

43. Laquer, *Confrontation*, p. 207.

44. OECD Oil Committee, Internal Files, 1973, IEA; Lantzke, "The OECD," in Vernon, ed., *The Oil Crisis*.

45. FEA, *U.S. Oil Companies*; OECD, *Crude Oil Import Prices: 1973–1980* (Paris, OECD, 1981); State Department, "OPEC."

46. Garratt, "Euro-American Energy Diplomacy," pp. 86–87.

47. Lantzke, "The OECD," in Vernon, ed. *The Oil Crisis*, p. 219.

48. Robert J. Lieber, *Oil and the Middle East War: Europe in the Energy Crisis* (Cambridge, Mass.: Harvard Center for International Affairs, 1976), p. 15.

49. United Nations, *World Energy Supplies: 1971-1975* (New York: United Nations, 1977), p. 101.

50. FEA, *U.S. Oil Companies*, p. 10.

51. Stobaugh, "The Oil Companies in the Crisis," in Vernon, ed., *The Oil Crisis*, p. 187; Robert Weiner, *The Oil Import Question in an International Context: Institutional and Economic Aspects of Consumer Cooperation*, Energy and Environmental Policy Center Working Paper E-81-06, Harvard University, June 1981, p. 7.

52. Office of the President, *International Economic Report of the President* (Washington: Government Printing Office, March 1976).

53. Ann-Margaret Walton, "Atlantic Bargaining over Energy," *International Affairs 52* (April 1976): 183.

54. Karl Kaiser, "The Energy Problem and Alliance Systems: Europe," *Adelphi Papers 115* (1975): 18.

55. Laquer, *Confrontation*, p. 213.

56. Henry Kissinger, *Years of Upheaval* (Boston: Little, Brown, 1982), p. 896; Lieber, *Oil and the Middle East War*; Louis Turner, "The European Community: Factors of Disintegration," *International Affairs 50* (July 1974): 404–415.

57. Congressional Quarterly, *Nixon: The Fifth Year of His Presidency* (Washington: *Congressional Quarterly,* 1974).

58. The quote is from Lieber, *Oil and the Middle East War*, p. 8; see also Laquer, *Confrontation*, p. 149, and Turner, "The European Community."

59. Romano Prodi and Alberto Clo, "Europe," in Vernon, ed., *The Oil Crisis*, p. 98; Jeffrey Record, *Revising U.S. Military Strategy* (McLean, Va.: Pergamon–Brassey, 1984), p. 61.

60. Quoted in Walton, "Atlantic Bargaining over Energy," p. 184.

61. Kissinger, *Years of Upheaval*, p. 901.

62. Walton, "Atlantic Bargaining," pp. 186–187.

63. Kissinger, *Years of Upheaval*, ch. 20, "Energy and the Democracies."

64. Ibid., p. 905.

65. Turner, "The European Community." On the conference, see also Lieber, "American Diplomatic Response."

66. Quoted in Kissinger, *Years of Upheaval*, p. 916.

67. Laquer, *Confrontation*, p. 180.

68. Walton, "Atlantic Bargaining," p. 189.

Chapter 8

1. Henry Kissinger, "Foreword," in Charles K. Ebinger, *The Critical Link* (Cambridge, Mass.: Ballinger, 1981), xx.

2. On the Soviet oil trade, see Arthur Jay Klinghoffer, *The Soviet Union and International Oil Politics* (New York: Columbia University Press, 1977).

3. On the IEA see Mason Willrich and Melvin Conant, "The International Energy Agency: An Interpretation and Assessment," *American Journal of International Law 71* (April 1977): 199–223; IEA, "The International Energy Agency," (Paris: IEA, 1974); Ulf Lantzke, "The OECD and Its International Energy Agency," in Vernon, ed., *The Oil Crisis*.

4. Willrich and Conant, "The International Energy Agency."

5. On supply rights see Robert Weiner, *The Oil Import Question in an International Context*, Energy and Environmental Policy Center Discussion Paper E-81-07, Harvard University, June 1981.

6. See Robert Keohane, "International Agencies and the Art of the Possible: The Case of the IEA," *Journal of Policy Analysis and Management 1* (Summer 1982): 469–481, who nevertheless makes an optimistic case for the agency.

7. Ebinger, *The Critical Link*, p. 119.

8. Adelman, *World Petroleum Market*, p. 273.

9. See Davis B. Bobrow and Robert T. Kudrle, "Energy R&D: In Tepid Pursuit of Collective Goods," *International Organization 33* (Spring 1979): 149–176; U.S. Congress, House Committee on Science and Technology, *International Cooperation in Energy Research and Development* (Washington: Government Printing Office, 1976).

10. Daniel Badger, "International Cooperation During Oil Supply Disruptions: The Role of the International Energy Agency," in George Horwich and David Leo

Weimer, eds., *Responding to International Oil Crises* (Washington: American Enterprise Institute, 1988), pp. 1-2.

11. For overviews of the market see Ebinger, *The Critical Link*, pp. 123–141; Ian Torrens, *Changing Structures in the World Oil Market* (Paris: Atlantic Institute for International Affairs, 1980); Ian Seymour, *OPEC: Instrument of Change* (London: Macmillan, 1980), pp. 126–147.

12. Seymour, *OPEC*, p. 148.

13. Torrens, *Changing Structures*, p. 29.

14. Schneider, *Oil Price Revolution*, p. 411.

15. Ibid.

16. U.S. Congress, Senate Committee on Energy and Natural Resources, *The Geopolitics of Oil* (Washington: Government Printing Office, 1980), p. 34.

17. G. John Ikenberry, "Market Solutions for State Problems: The International and Domestic Politics of American Oil Decontrol," *International Organization 42* (Winter 1988), p. 169.

18. U.S. Central Intelligence Agency, *The International Energy Situation: Outlook to 1985* (Washington: Government Printing Office, 1977); Dankwart Rustow, "U.S.–Saudi Relations and the Oil Crises of the 1980s," *Foreign Affairs 55* (April 1977), p. 494.

19. Terisa Turner, "Iranian Oilworkers in the 1978–79 Revolution," in *Oil and Class Struggle*.

20. Richard W. Cottam, "Iran and the Middle East," in Spiegel, ed., *The Middle East and the Western Alliance*, p. 215.

21. "Spot Product Princes Hit 'Incredible' Levels in Europe," *Petroleum Intelligence Weekly*, November 13, 1978.

22. See *PIW* issues for the period.

23. "No Quick Trigger of Oil Sharing Seen by IEA," *PIW*, November 13, 1978.

24. Richard N. Cooper, International Economics Seminar, Harvard University, April 20, 1983.

25. "Exxon Implements IEA Concept to Allocate Oil Supply," *PIW*, February 26, 1979.

26. See Keohane, "International Agencies and the Art of the Possible;" U.S. Department of Energy, *Energy Security in the Industrial Nations* (Washington: Department of Energy, January 1983); Richard Bissell, "The West in Concert," in *Oil Diplomacy* (Philadelphia: Foreign Policy Research Institute, 1980).

27. Bissell, "The West in Concert," p. 76.

28. *PIW*, April 30, 1979.

29. Richard Mancke, "The American Response: On the Job Training?" in *Oil Diplomacy*, p. 33; Joseph A. Yager, "The Energy Battles of 1979," in Goodwin, *Energy Policy in Perspective*, pp. 606–607.

30. Ibid; *PIW*, May 21, 1979.

31. Bissell, "The West in Concert," p. 76.

32. Weiner, *The Oil Import Question*.

33. *PIW*, July 2, 1979.

34. Daniel Badger and Robert Belgrave, *Oil Supply and Price: What Went Right in 1980?* (London: British Institutes' Joint Energy Policy Programme, 1982), p. 95.

35. Weiner, *The Oil Import Question*, p. 67.

36. Kupchan, *Persian Gulf and the West*, p. 70.

37. Jimmy Carter, *Keeping Faith* (London: Colliers, 1982), p. 92.

38. Harold H. Saunders, "The Iran–Iraq War: Implications for US Policy," in Thomas Naff, ed., *Gulf Security and the Iran–Iraq War* (Washington: National Defense University Press, 1985), p. 62.

39. See Kupchan, *Persian Gulf and the West*, pp. 178–179.

40. On the Carter Doctrine see Frans R. Bax, "Energy Security in the 1980s: The Response of U.S. Allies," in Donald J. Goldstein, ed., *Energy and National Security* (Washington: National Defense University Press, 1981), pp. 34–45; Walter J. Levy, "Oil and the Decline of the West," *Foreign Affairs 58* (Summer 1980): 999–1015; Kupchan, *Persian Gulf and the West*, pp. 83–98. For an article that places the Carter Doctrine in historical perspective, see Melvyn P. Leffler, "From the Truman Doctrine to the Carter Doctrine," *Diplomatic History 7* (Fall 1983): 245–266.

41. On U.S. force planning, see Congressional Research Service, *Western Vulnerability to a Disruption of Persian Gulf Oil Supplies: U.S. Interests and Options* (Washington: Congressional Research Service, March 24, 1983).

42. Bax, "Energy Security," p. 37.

43. Quoted in Kupchan, *Persian Gulf and the West*, p. 183.

44. Quoted in Bax, "Energy Security," p. 37.

45. Sumner Benson, "The Defense Department and the Politics of Economic Security," March 11, 1986, mimeograph.

46. Badger and Belgrave, "What Went Right?" p. 118.

47. DOE, *Energy Security*, p. 27.

48. *PIW*, February 25, 1980.

49. *PIW*, April 28, 1980.

50. DOE, *Energy Security*, p. 27.

51. U.S. Congress, Senate, *The Geopolitics of Oil*, pp. 17–19; Mark A. Heller, *The Iran–Iraq War: Implications for Third Parties* (Tel Aviv: Jaffee Center for Strategic Studies, 1984).

52. DOE, *Energy Security*, p. 28; Rodney T. Smith, "International Energy Cooperation: The Mismatch Between IEA Policy Actions and Policy Goals," in Horwich and Weimer, *Responding*, p. 37; Robert Keohane, "International Agencies and the Art of the Possible: The Case of the IEA," *Journal of Policy Analysis and Management 1* (1982): 469–481; Badger, "What Went Right?" p. 120.

53. Ibid.

54. Badger, "What Went Right?" p. 121.

55. DOE, *Energy Security*, p. 61.

56. U.S. Congress, Senate, *Geopolitics of Oil*, p. 32.

57. See DOE, *Energy Security*; Smith, "Mismatch;" Badger, "What Went Right?"; Keohane, "International Agencies."

Chapter 9

1. Raymond Aron, *Peace and War* (Garden City, N.Y.: Doubleday, 1968), p. 54.

2. See Lincoln Gordon, "Economic Aspects of Coalition Diplomacy: The NATO Experience," *International Organization 10* (Autumn 1956): 529–543.

3. Robert Keohane, "Hegemonic Leadership and U.S. Foreign Economic Policy in the 'Long Decade' of the 1950s," in William P. Avery and David P. Rapkin, eds., *America in a Changing World Political Economy* (New York: Longman, 1982), p. 53.

4. John P. Weyant, "Coordinated Stock Drawdowns: Pros and Cons," in Horwich and Weimar, *Responding to International Oil Crises,* p. 182.

5. Adelman, *World Petroleum Market,* p. 273.

6. Ibid., p. 267.

7. Harold Lubell, "Security of Supply and Energy Policy in Western Europe," *World Politics 13* (April 1961): 401–422.

8. For a hopeful view of the IEA, see Keohane, *After Hegemony.*

9. U.S. Department of Energy, Energy Security (Washington, D.C.: U.S. Department of Energy, 1987), pp. 231–234.

10. Michael Don Ward, "Research Gaps in Alliance Dynamics," *Monograph Series in World Affairs 19* (Book 1, 1982), p. 55.

11. Oran Young, *The Politics of Force* (Princeton, N.J.: Princeton University Press, 1968), pp. vii–viii.

12. But see Stephen M. Walt, *The Origins of Alliances* (Ithaca, N.Y.: Cornell University Press, 1987).

13. Hedley Bull, "European Self-Reliance and the Reform of NATO," *Foreign Affairs 61* (Spring 1983), p. 876.

14. Ibid., p. 880.

15. See Bruce Russett, "The Mysterious Case of Vanishing Hegemony," *International Organization 39* (Spring 1985): 207–231.

Selected Bibliography

Archival Sources

National Archives, Washington, D.C.

Records of the Agency for International Development, Record Group (RG) 286.
Records of Allied Operational and Occupation Headquarters, RG 331.
Records of the Bipartite Control Office, RG 260.
Records of the Combined Boards, RG 179.
Records of the Department of State, RG 59.
Records of Foreign Service Posts, RG 84.
Records of Interdepartmental Committees, RG 353.
Records of International Conferences, RG 43.
Records of the War Department General and Special Staffs, RG 165.

Harry S. Truman Library, Independence, Mo.

Dean Acheson Papers.
William Clayton Papers.
Charles P. Kindleberger Oral History Interview.
James Riddleberger Oral History Interview.

Organization for Economic Cooperation and Development, Paris, France

Records of the Organization for European Economic Cooperation Oil Committee.

Records of the Organization for Economic Cooperation and Development Oil Committee.
Records of the International Energy Agency.

U.S. Department of Energy Archives, Germantown, Md.

Records of the Secretariat, 1951–1958.

Published Government and Official Documents

Commission of the European Communities. *Medium Term Prospects and Guidelines in the Community Gas Sector*. Brussels, 1972.

Committee for European Economic Cooperation. *General Report*. Washington: Department of State, 1947.

———. *Technical Reports*. Washington: Department of State, 1947.

European Coal Organization. *The European Coal Organization*. London, 1947.

International Energy Agency. *Natural Gas: Prospects to 2000*. Paris, 1982.

———. *The International Energy Agency*. Paris, 1974.

———. *World Energy Outlook*. Paris, 1982.

Organization for Economic Cooperation and Development. *From Marshall Plan to Global Interdependence*. Paris, 1978.

Organization for European Economic Cooperation. *Europe's Need for Oil: Implications and Lessons of the Suez Crisis*. Paris, 1958.

———. *The Coal Industry in Europe*. Paris, 1954.

U.S. Bureau of the Budget. *The United States and War*. Washington: Government Printing Office, 1946.

U.S. Cabinet Task Force on Oil Import Controls. *The Oil Import Question*. Washington: Government Printing Office, 1970.

U.S. Congress, House Committee on Interstate and Foreign Commerce, *Fuel Investigation—Petroleum and the European Recovery Program*. H.R. 1438, 80th Cong., 2d sess., 1948.

U.S. Congress, Senate Committee on Foreign Relations. *Multinational Corporations and United States Foreign Policy: Part 7*. 93rd Cong., 2d sess., 1974.

U.S. Department of Commerce. *Foreign Aid: 1940–1951*. Washington: Government Printing Office, 1952.

U.S. Department of State. *Energy Resources of the World*. Washington: Government Printing Office, 1949.

U.S. Economic Cooperation Administration. *Coal and Related Solid Fu-*

els Commodity Study. Washington: Government Printing Office, 1949.

Books, Articles, and Dissertations

Acheson, Dean. *Present at the Creation.* New York: Norton, 1969.

Adelman, M. A. *The World Petroleum Market.* Baltimore: Johns Hopkins University Press, 1972.

Ambrose, Stephen E. *Rise to Globalism.* New York: Penguin Books, 1983.

Anderson, Irvine. *Aramco, the United States, and Saudi Arabia.* Princeton, N.J.: Princeton University Press, 1981.

Arkes, Hadley. *Bureaucracy, the Marshall Plan, and the National Interest.* Princeton, N.J.: Princeton University Press, 1972.

Aron, Raymond. *Peace and War.* New York: Doubleday, 1968.

Ashley, Richard K. *The Political Economy of War and Peace.* New York: Nichols, 1980.

Aubrey, Henry G. *Atlantic Economic Cooperation.* New York: Praeger, 1967.

Avery, William, and Rapkin, David, eds. *America in a Changing World Political Economy.* New York: Longmans, 1982.

Balabkins, Nicholas. *Germany Under Direct Controls.* New Brunswick, N.J.: Rutgers University Press, 1964.

Baldwin, David. *Economic Statecraft.* Princeton, N.J.: Princeton University Press, 1985.

Bareau, Paul. *The Sterling Area.* London: Longmans Green & Co., 1948.

Beloff, Max. *The United States and the Unity of Europe.* Washington: Brookings Institution, 1963.

Bergsten, C. Fred; Horst, Thomas; and Moran, Theodore. *American Multinationals and American Interests.* Washington: Brookings Institution, 1978.

Bernstein, Barton J., ed. *Politics and Policies of the Truman Administration.* Chicago: Quadrangle Books, 1972.

Blair, John. *The Control of Oil.* New York: Pantheon, 1976.

Bobrow, Davis B., and Kurdle, Robert T. "Energy R&D: In Tepid Pursuit of Collective Goods." *International Organization 33* (Spring 1979): 149–175.

Bohi, Douglas, and Russell, Milton. *Limiting Oil Imports.* Baltimore: Johns Hopkins University Press, 1978.

Bowie, Robert R. *Suez 1956*. New York: Oxford University Press, 1974.

Brookings Institution. *Major Problems of United States Foreign Policy*. Washington: Brookings Institution, 1949.

———. *Anglo–American Economic Relations*. Washington: Brookings Institution, 1950.

Brown, Seyom. *New Forces in World Politics*. Washington: Brookings Institution, 1974.

Brown, William Adams Jr., and Opie, Redvers. *American Foreign Assistance*. Washington: Brookings Institution, 1953.

Buchanan, James M. *The Demand and Supply of Public Goods*. Chicago: Rand McNally, 1965.

Bucknell, Howard. *Energy and National Defense*. Lexington, Ky.: University Press of Kentucky, 1981.

Bull, Hedley. "European Self-Reliance and the Reform of NATO." *Foreign Affairs 61* (Spring 1983): 874–892.

Calleo, David, and Rowland, Benjamin. *America and the World Political Economy*. Bloomington: Indiana University Press, 1977.

Campbell, John C. *Defense of the Middle East*. New York: Praeger, 1960.

Chester, Edward W. *United States Oil Policy and Diplomacy*. Westport, Conn.: Greenwood Press, 1983.

Clark, John G. *Energy and The Federal Government*. Urbana, Ill.: University of Illinois Press, 1987.

Claude, Inis. *Swords into Plowshares*. New York: Random House, 1971.

Cohen, Benjamin J. *Organizing the World's Money*. New York: Basic Books, 1971.

Cooper, Chester L. *The Lion's Last Roar: Suez 1956*. New York: Harper & Row, 1978.

Cooper, Richard N. "Trade Policy Is Foreign Policy." *Foreign Policy 9* (Winter 1972–73): 46–61.

Cottam, Richard. *Nationalism in Iran*. Pittsburgh: University of Pittsburgh Press, 1964.

Cowhey, Peter F. *The Problems of Plenty*. Berkeley: University of California Press, 1985.

Dam, Kenneth W. "Implementation of Import Quotas: The Case of Oil." *Journal of Law and Economics 14* (April 1971): 1–60.

Deese, David. "Oil, War, and Grand Strategy." *Orbis* (Fall 1981): 525–555.

DeNovo, John A., *American Interests and Policies in the Middle East*. Minneapolis: University of Minnesota Press, 1963.

Deporte, A. W. *DeGaulle's Foreign Policy: 1944–1946*. Cambridge, Mass.: Harvard University Press, 1968.

———. *Europe Between the Superpowers*. New Haven, Conn.: Yale University Press, 1979.

Dewhurst, J. Frederic. *Europe's Needs and Resources.* New York: Twentieth Century Fund, 1961.

Diebold, William Jr. *The Schuman Plan.* New York: Praeger, 1959.

———. *The United States and the Industrial World.* New York: Praeger, 1972.

Donnison, F. S. V. *Civil Affairs and Military Government: North-West Europe, 1944–1946.* London: HMSO, 1961.

Doran, Charles F. *Myth, Oil, and Politics.* New York: Free Press, 1977.

Dougherty, James E., and Pfaltzgraff, Robert Jr. *Contending Theories of International Relations.* New York: Harper & Row, 1981.

Ebinger, Charles K. *The Critical Link: Energy and National Security in the 1980s.* Cambridge, Mass.: Ballinger, 1982.

Eckes, Alfred E. Jr. *The United States and the Global Struggle for Minerals.* Austin: University of Texas Press, 1979.

Fanning, Leonard. *Foreign Oil and the Free World.* New York: McGraw-Hill, 1954.

Foreign Policy Research Institute. *Oil Diplomacy.* Philadelphia: Foreign Policy Research Institute, 1980.

Fremont, Jacques. *The Saar Conflict.* New York: Praeger, 1960.

Frey, Bruno. *Modern Political Economy.* New York: John Wiley & Sons, 1978.

Frost, Ellen L., and Stent, Angela E. "NATO's Troubles with East–West Trade." *International Security 8* (Summer 1983): 179–200.

Gaddis, John Lewis. *Strategies of Containment.* New York: Oxford University Press, 1982.

———. *The United States and the Origins of the Cold War: 1941–1947.* New York: Columbia University Press, 1972.

Gardner, Lloyd. *Economic Aspects of New Deal Diplomacy.* Madison: University of Wisconsin Press, 1964.

Gardner, Richard N. *Sterling–Dollar Diplomacy.* New York: McGraw-Hill, 1969.

Gilpin, Robert. *U.S. Power and the Multinational Corporation.* New York: Basic Books, 1975.

Gimbel, John. *The Origins of the Marshall Plan.* Stanford, Calif.: Stanford University Press, 1975.

Goldstein, Donald J., ed. *Energy and National Security.* Washington: National Defense University Press, 1981.

Goodwin, Craufurd, ed. *Energy Policy in Perspective.* Washington: Brookings Institution, 1981.

Gordon, Lincoln. "Economic Aspects of Coalition Diplomacy—The NATO Experience." *International Organization 10* (Autumn 1956): 530–543.

Gourevitch, Peter. "The Second Image Reversed: The International Sources of Domestic Politics." *International Organization 32* (Autumn 1978): 881–911.

Gowa, Joanne. "Hegemons, IOs, and Markets: The Case of the Substitution Account." *International Organization 38:* 661–683.

Gowing, Margaret. *Independence and Deterrence: Britain and Atomic Energy, 1945–1952.* London: Macmillan, 1974.

Griffin, A.R. *The British Coalmining Industry.* Buxton, U.K.: Moorland, 1977.

Groen, E. "The Significance of the Marshall Plan for the Petroleum Industry in Europe—Historical Review of the Period 1947–1950." *Report of the Third World Petroleum Congress.* The Hague: n.p., 1951.

Hager, Wolfgang. *Europe's Economic Security.* Paris: Atlantic Institute for International Affairs, 1975.

Hanreider, Wolfram. *West German Foreign Policy: 1949–1963.* Stanford, Calif.: Stanford University Press, 1963.

Harris, Seymour E., ed. *Foreign Economic Policy for the United States.* Cambridge, Mass.: Harvard University Press, 1948.

Haviland, H. Field Jr., and Wilcox, Francis O., eds. *The Atlantic Community: Progress and Prospects.* New York: Praeger, 1963.

Haynes, William. *Nationalization in Practice: The British Coal Industry.* Boston: Harvard Business School Press, 1953.

Hewlett, Richard G. and Jack M. Holl; *Atoms for Peace and War: Eisenhower and the Atomic Energy Commission, 1953–1960.* Berkeley: University of California Press, 1989.

Hirschman, Albert O. *National Power and the Structure of Foreign Trade.* Berkeley: University of California Press, 1945.

Hogan, Michael. *The Marshall Plan.* New York: Cambridge University Press, 1988.

Howard, Michael. *Security Issues and Western Europe's Political Future.* New York: Trilateral Commission, 1984.

Issawi, Charles, and Yeganeh, Mohammed. *The Economics of Middle East Oil.* New York: Praeger, 1962.

Jackson, Scott. "Prologue to the Marshall Plan: The Origins of the American Commitment for a European Recovery Program." *Journal of American History* (March 1979): 1043–1068.

Jentleson, Bruce. "From Consensus to Conflict: The Domestic Political Economy of East–West Trade Policy." *International Organization 38* (Autumn 1984): 625–660.

Kaiser, Karl. *The Energy Problem and Alliance Systems: Europe.* London: International Institute for Strategic Studies, 1975.

Kalt, Joseph. *The Economics and Politics of Oil Price Regulation.* Cambridge, Mass.: MIT Press, 1981.

Kapstein, Ethan B. "Alliance Energy Security: 1945–1983." *Fletcher Forum* (Winter 1984): 91–116.

———. "The Marshall Plan and Industrial Policy." *Challenge* (May/June 1984): 55–59.

Katzenstein, Peter, ed. *Between Power and Plenty.* Madison: University of Wisconsin Press, 1978.

Keohane, Robert O. *After Hegemony.* Princeton, N.J.: Princeton University Press, 1984.

———. "International Agencies and the Art of the Possible: The Case of the IEA." *Journal of Policy Analysis and Management 1* (Summer 1982): 469–481.

———. "The International Energy Agency: State Influence and Transgovernmental Politics." *International Organization 32* (Autumn 1978): 929–951.

———. "State Power and Industry Influence: American Foreign Oil Policy in the 1940s." *International Organization 36* (Winter 1982): 165–183.

Keohane, Robert O., and Nye, Joseph S. *Power and Interdependence.* Boston: Little, Brown, 1977.

Keohane, Robert O., and Nye, Joseph S., eds. *Transnational Relations and World Politics.* Cambridge, Mass.: Harvard University Press, 1972.

Kindleberger, Charles. *The Dollar Shortage.* Cambridge, Mass.: MIT Press, 1950.

———. "Toward the Marshall Plan: A Memoir of Policy Development, 1945–1947." (mimeograph.)

Kissinger, Henry. *Years of Upheaval.* Boston: Little, Brown, 1982.

Klein, Burton H. "Germany's Economic Preparations for War." Ph.D. Dissertation, Harvard University, 1952.

Klinghoffer, Arthur Jay. *The Soviet Union and International Oil Politics.* New York: Columbia University Press, 1977.

Kohl, Wilfrid L., ed. *After the Second Oil Crisis.* Lexington, Mass.: Lexington Books, 1982.

Kolko, Gabriel. *The Roots of American Foreign Policy.* Boston: Beacon Press, 1969.

Kolko, Gabriel, and Kolko, Joyce. *The Limits of Power.* New York: Harper & Row, 1972.

Kosnik, Joseph T. *Natural Gas Imports from the Soviet Union: Financing the North Star Joint Venture Project.* New York: Praeger, 1975.

Kovrig, Bennett. *The Myth of Liberation.* Baltimore: Johns Hopkins University Press, 1973.

Kramish, Arnold. *The Peaceful Atom in Foreign Policy.* New York: Harper and Row, 1963.

Krasner, Stephen D. "A Statist Interpretation of American Oil Policy Toward the Middle East." *Political Science Quarterly 94* (Spring 1979): 77–96.

———. *Defending the National Interest: Raw Materials Investments and U.S. Foreign Policy.* Princeton, N.J.: Princeton University Press, 1978.

———, ed. *International Regimes.* Ithaca, N.Y.: Cornell University Press, 1983.

Kuklick, Bruce. *American Policy and the Division of Germany.* Ithaca, N.Y.: Cornell University Press, 1972.

Kuniholm, Bruce R. *The Origins of the Cold War in the Near East.* Princeton, N.J.: Princeton University Press, 1980.

LaFeber, Walter. *America, Russia, and the Cold War: 1945–1971.* New York: John Wiley & Sons, 1972.

Laquer, Walter. *The Road to War.* Harmondsworth, U.K.: Penguin Books, 1968.

Leeman, Wayne. *The Price of Middle East Oil.* Ithaca, N.Y.: Cornell University Press, 1962.

Leffler, Melvyn P. "The United States and the Strategic Dimensions of the Marshall Plan." *Diplomatic History 12* (Summer 1986): 277–306.

Lieber, Robert. *Oil and the Middle East War: Europe in the Energy Crisis.* Cambridge, Mass.: Harvard University Center for International Affairs, 1976.

Lindblom, Charles E. *Politics and Markets.* New York: Basic Books, 1977.

Liska, George. *Nations in Alliance.* Baltimore: Johns Hopkins University Press, 1962.

Lister, Louis. *Europe's Coal and Steel Community.* New York: Twentieth Century Fund, 1960.

Longrigg, Stephen. *Oil in the Middle East.* New York: Oxford University Press, 1954.

Lubell, Harold, "Security of Supply and Energy Policy in Western Europe," *World Politics 13* (April 1961: 401–422)

Lucas, N. J. D. *Energy and the European Communities.* London: Europa, 1977.

Lundestad, Geir. *The American Non-Policy Towards Eastern Europe.* New York: Humanities Press, 1975.

Maier, Charles S. "The Two Postwar Eras and the Conditions for Stability in Twentieth Century Western Europe." *American Historical Review 86* (April 1981): 327–352.

Mason, E. S. "American Security and Access to Raw Materials." *World Politics 1* (January 1949): 147–160.

Maull, Hanns. *Europe and World Energy*. London: Butterworths, 1980.

Mendershausen, Horst. *Coping with the Oil Crisis*. Baltimore: Johns Hopkins University Press, 1976.

———. "Dollar Shortage and Oil Surplus in 1949–50." *Essays in International Finance 11* (November 1959).

Miliband, Ralph. *The State in Capitalist Society*. London: Weidenfeld & Nicolson, 1969.

Miller, Aaron David. *Search for Security*. Chapel Hill, N.C.: University of North Carolina Press, 1980.

Millis, Walter, ed. *The Forrestal Diaries*. New York: Viking, 1951.

Milward, Alan S. *The Reconstruction of Western Europe: 1945–1951*. London: Methuen, 1984.

Morgenthau, Hans. *Power Among Nations*. New York: Knopf, 1968.

Nye, Joseph S. "Collective Economic Security." *International Affairs 50* (October 1974): 584–598.

———. "Energy and Security in the 1980s." *World Politics 35* (October 1982): 121–134.

Odell, Peter. *Oil and World Power*. Harmondsworth, U.K.: Penguin Books, 1972.

Olson, Mancur. *The Logic of Collective Action*. Cambridge, Mass.: Harvard University Press, 1965.

Olson, Mancur, and Zeckhauser, Richard. "An Economic Theory of Alliances." *Review of Economics and Statistics 48* (August 1966): 266–279.

Oppenheimer, Joe. "Collective Goods and Alliances: A Reassessment." *Journal of Conflict Resolution 23* (September 1979): 387–407.

Painter, David S. *Oil and the American Century*. Baltimore: Johns Hopkins University Press, 1986.

Passarelli, Antonio. "Energy Policies in the European Community: 1950–1972." Ph.D. Dissertation, University of Notre Dame, 1975.

Penrose, Edith. *The International Oil Industry in the Middle East*. Cairo: National Bank of Egypt, 1968.

———. *The Large International Firm in Developing Countries*. London: George Allen & Unwin, 1968.

Pollard, Robert A. *Economic Security and the Origins of the Cold War*. New York: Columbia University Press, 1985.

Pounds, Norman J.G. *The Ruhr*. Bloomington: Indiana University Press, 1952.

Price, Harry. *The Marshall Plan and Its Meaning.* Ithaca, N.Y.: Cornell University Press, 1955.

Rappoport, Armin. "The United States and European Integration: The First Phase." *Diplomatic History 5* (Spring 1981): 121–149.

Record, Jeffrey. *Revising U.S. Military Strategy.* McLean, Va.: Pergamon-Brassey, 1984.

Reichart, John, and Sturm, Steven, eds. *American Defense Policy.* Baltimore: Johns Hopkins University Press, 1982.

Resources for the Future. *U.S. Energy Policies: An Agenda for Research.* Baltimore: Johns Hopkins University Press, 1968.

Reuter, Paul. *Organisations Européennes.* Paris: PUF, 1965.

Ritchie, Ronald S. *NATO: The Economics of an Alliance* (Toronto: The Ryerson Press, 1956).

Rogers, John M. "The International Authority for the Ruhr." Ph.D. Dissertation, American University, Washington, D.C. 1960.

Roosevelt, Kermit. *Countercoup.* New York: McGraw-Hill, 1979.

Rostow, W. W. *The World Economy.* Austin: University of Texas Press, 1978.

Ruggie, John G. "Collective Goods and Future International Collaboration." *American Political Science Review 66* (September 1972): 874–893.

Russett, Bruce. "Dimensions of Resource Dependence." *International Organization 38* (Summer 1984): 482–499.

———. "Security and the Resources Scramble: Will 1984 Be Like 1914?" *International Affairs 58* (Winter 1981/82): 42–58.

Russett, Bruce, and Sullivan, John. "Collective Goods and International Organization." *International Organization 25* (Autumn 1971): 845–865.

Sampson, Anthony. *The Seven Sisters.* New York: Bantam Books, 1976.

Samuels, Nathaniel. "The European Coal Organization." *Foreign Affairs 26* (October 1948): 728–736.

Saunders, Richard. "Military Force in the Foreign Policy of the Eisenhower Presidency." *Political Science Quarterly 100* (Spring 1985): 96–116.

Schlesinger, Arthur Jr., ed. *The Dynamics of World Power.* New York: Chelsea House, 1973.

Schlesinger, James. *The Political Economy of National Defense.* New York: Praeger, 1960.

Schmidt, Hubert G. *Policy and Functioning in Industry.* Karlsruhe, West Germany: U.S. Military Government in Germany, 1950.

Schneider, Steven. *The Oil Price Revolution.* Baltimore: Johns Hopkins University Press, 1984.

Scott, John. *The Soviet Economic Offensive*. New York: Time, Inc., 1961.

Sherwin, Martin. *A World Destroyed*. New York: Vintage Books, 1977.

Shonfield, Andrew. *Modern Capitalism*. London: Oxford University Press, 1965.

Smith, Jean Edward, ed. *The Papers of General Lucius D. Clay*. Bloomington: Indiana University Press, 1974.

Snidal, Duncan. "The Limits of Hegemonic Stability Theory." *International Organization 39* (Autumn 1985): 579–614.

Solomon, Robert. *The International Monetary System*. New York: Harper & Row, 1977.

Spykman, Nicholas J. *America's Strategy in World Politics*. New York: Harcourt Brace, 1942.

Stein, Jonathan. *The Soviet Bloc, Energy and Western Security*. Lexington, Mass.: Lexington Books, 1983.

Stent, Angela E. *Soviet Energy and Western Europe*. New York: Praeger, 1982.

Stivers, William. *Supremacy and Oil*. Ithaca, N.Y.: Cornell University Press, 1982.

———. *America's Confrontation with Revolutionary Change in the Middle East, 1945–83*. London: Macmillan Press, 1986.

Stoff, Michael B. *Oil, War, and American Security*. New Haven, Conn.: Yale University Press, 1980.

Strange, Susan. "What Is Economic Power, and Who Has It?" *International Journal 30* (Spring 1975): 207–224.

Tanzer, Michael. *The Political Economy of Oil and the Underdeveloped Countries*. Boston: Beacon Press, 1969.

Thomas, Hugh. *Suez*. New York: Harper & Row, 1967.

Torrens, Ian. *Changing Structures in the World Oil Market*. Paris: Atlantic Institute for International Affairs, 1980.

Turner, Louis. "The European Community: Factors of Disintegration." *International Affairs 50* (July 1974): 404–415.

———. *The Oil Companies in the International System*. London: George Allen & Unwin, 1978.

Vernon, Raymond, ed. *The Oil Crisis*. New York: Norton, 1976.

Vietor, Richard K. *Energy Policy in America Since 1945*. New York: Cambridge University Press, 1984.

Walt, Stephen M. *The Origins of Alliances*. Ithaca, N.Y.: Cornell University Press, 1987.

Wandycz, Pietor S. *The United States and Poland*. Cambridge, Mass.: Harvard University Press, 1980.

Weiner, Robert. *The Oil Import Question in an International Context*.

Cambridge, Mass.: Harvard University Kennedy School of Government, 1981.

Wightman, David. *Economic Cooperation in Europe.* New York: Praeger, 1956.

Wilcox, Clair. *Public Policies Toward Business.* Homewood, Ill.: Irwin, 1971.

Wolfers, Arnold, ed. *Alliance Policy in the Cold War.* Baltimore: Johns Hopkins University Press, 1959.

Yergin, Daniel. *Shattered Peace.* Boston: Houghton Mifflin Co., 1977.

Young, Oran. *Resource Regimes.* Berkeley: University of California Press, 1982.

Zupnick, Elliot. *Britain's Postwar Dollar Problem.* New York: Columbia University Press, 1954.

Index